Fast Track Ham Radio Facts

A collection of useful knowledge for informed amateur radio operators.

Compiled and edited by Michael Burnette, AF7KB

All original content © Copyright 2019, 2020 by Michael Burnette. All rights reserved.

Original content may not be reproduced or used in any manner whatsoever without the express written permission of the publisher except for the use of brief quotations in a book review.

Corrections? Suggestions?
AF7KB@fasttrackham.com

Table of Contents

The Radio Amateur's Code 9

Your License Privileges 10
    Technician 10
    General 10
    Advanced 11
    Amateur Extra 11

Frequency Allocations & Band Plans 12
    Super Low Frequency (SLF) 12
        2200 Meters 135.7 - 137.8 kHz 12
    Ultra-Low Frequency (ULF) 13
        630 Meters 472 kHz – 479 kHz 13
    Medium Frequency (MF) 13
        160 Meters 1.8 – 2.0 MHz 13
    High Frequency 14
        80 Meters 3.5 – 4.0 MHz 14
        60 Meters 5.3 MHz, Channelized 15
        40 Meters 7.0 MHz – 7.3 MHz 16
        30 Meters 10.100 – 10.150 MHz 17
        20 Meters 14.000 – 14.350 MHz 17
        17 Meters 18.068 – 18.168 MHz 18
        15 Meters 21.000 – 21.450 MHz 18
        12 Meters 24.890 – 24.990 MHz 19
        10 Meters 28.000 – 29.700 MHz 19
    Very High Frequencies (VHF) 20
        6 Meters 50.000 – 54.000 MHz 20
        2 Meters 144.000 – 148.000 MHz 21
        1.25 Meters 219.000 – 225.000 MHz 22
    Ultra High Frequencies (UHF) 23
        70 Centimeters 420.000 – 450.000 MHz 23
        33 Centimeters 24
        23 Centimeters 1240.000 – 1300.000 MHz 25
        13 Centimeters 2300.000 – 2450.000 MHz 26
    Super High Frequency (SHF) 27
        9 Centimeters 3300.000 – 3500.000 MHz 27
        5 Centimeters 5650.000 – 5925.000 MHz 30
        3 Centimeters 10000.000 – 10050.000 MHz 31
        Above 10.50 GHz 32

Popular HF & 6-meter Digital Mode Frequencies 32

Family Radio Service and General Mobile Radio Service Frequencies 34

MURS (Multi-Use Radio Service) Frequencies 35

NOAA Weather Radio Frequencies 35

Citizens Band Frequencies 35

Aircraft VHF Frequencies 36

Marine VHF Frequencies	37
NCDXF Beacon Stations	39
NATO Phonetic Alphabet	40
International Morse Code	40
Call Sign Formats	41
US Call Sign Regions	41
Allocations of World Call Sign Prefixes	42
World Call Sign Prefixes by Allocation	60
Third Party Operating Agreements	73
World Time Chart	74
World Time by Country	75
RST Signal Report Guidelines	85
Q Signals	86
Prosigns	88
ARRL QN Signals for CW Net Use	90
ARRL Prosigns For CW Net Use	91
ITU Emission Types	93
Common Ham Radio Emission Types	95
Wire Size and Ampacity Guidelines	96
Wire Types	97
Electrical Wire Run Lengths vs. Voltage Drop	98
Coaxial Cable Specifications – Power Capacity	99
Coaxial Cable Specifications - Attenuation	100
Schematic Symbols	101
Antenna Lengths	103
Optimum "Random" Wire Antenna Lengths	104
Power Loss from SWR	105
RF Exposure Limits	106
World Voltages	106
Electrical Values & Abbreviations	107
Metric Electronic Values	108
Resistor Color Codes	109
Mica Capacitor Color Code	110
Ohm's Law	111
Electronic Formulas	112

- Series Resistance, Series Inductance, and Parallel Capacitance ... 112
- Parallel Resistance, Parallel Inductance, and Series Capacitance ... 112
- Convert Frequency to Wavelength ... 112
- Convert Wavelength to Frequency ... 112
- Antenna Length, ½ Wave Antenna ... 112
- Adjusting Length of Dipole Antenna ... 112
- Power Loss from SWR ... 112
- Antenna Length, ¼ Wave Antenna ... 112
- Percent Change of Power Increase from Decibels ... 112
- Decibels from Power Change ... 112
- Percent Power Remaining After dB Change ... 112
- dB to Power Ratio ... 112
- Power Ratio to dB ... 112
- Intermodulation Frequency ... 112
- Resonant Frequency of a Circuit ... 112
- Inductive Reactance ... 112
- Capacitive Reactance ... 112
- Impedance (Series) ... 112
- Impedance (Parallel) ... 112
- Q of a Series Resonant Circuit ... 112
- Q of a Parallel Resonant Circuit ... 112
- Half power bandwidth ... 113
- Time Constant of a Capacitor ... 113
- Time Constant of an Inductor ... 113
- Susceptance ... 113
- Phase Angle ... 113
- Power Factor ... 113
- RMS Voltage from Peak Voltage ... 113
- Peak Voltage from RMS Voltage ... 113
- Peak Voltage from Peak-to-Peak ... 113
- Turns on a Coil for a Desired Inductance ... 113
- Voltage at the Secondary Winding of a Transformer Based on Turns (N) ... 113
- Power Dissipation of a Series Connected Linear Voltage Regulator ... 113
- Gain of an Op-amp ... 113
- Modulation Index (Deviation ratio) ... 113
- Bandwidth from WPM or Baud rate ... 113
- Antenna Efficiency ... 114
- Impedance Matching Transformer ... 114
- Bandwidth of an FM Signal (Carrying multiple frequencies.) ... 114
- Calculate Radius of Fresnel Zones ... 114
- Calculate Critical Frequency ... 114

Space Weather Terms ... 115

Sunspots and Sunspot Numbers ... 115

The Solar Flux Index ... 115

K-index ... 115

A-index ... 115

Electron Flux ... 115

Proton Flux ... 116

Solar Wind ... 116

X-ray Flux ... 116

304A ... 116

B_z ... 116

Checking the Space Weather... 116
FCC Certified VEC's 117
Smith Chart 118
FCC Rules & Regulations Part 97—Amateur Radio Service 119
Subpart A—General Provisions 119
§97.1 Basis and purpose. 119
§97.3 Definitions. 119
§97.5 Station license required. 122
§97.7 Control operator required. 123
§97.9 Operator license grant. 123
§97.11 Stations aboard ships or aircraft. 123
§97.13 Restrictions on station location. 123
Table 1 to § 1.1310(e)(1) - Limits for Maximum Permissible Exposure (MPE) 124
§ 97.15 Station antenna structures. 124
§97.17 Application for new license grant. 124
§97.19 Application for a vanity call sign. 125
§97.21 Application for a modified or renewed license grant. 126
§97.23 Mailing address. 126
§97.25 License term. 127
§97.27 FCC modification of station license grant. 127
§97.29 Replacement license grant document. 127
§97.31 Cancellation on account of the licensee's death. 127
Subpart B—Station Operation Standards. 127
§97.101 General standards. 127
§97.103 Station licensee responsibilities. 127
§97.105 Control operator duties. 128
§97.107 Reciprocal operating authority. 129
§97.109 Station control. 129
§97.111 Authorized transmissions. 129
§97.113 Prohibited transmissions. 130
§97.115 Third party communications. 130
§97.117 International communications. 131
§97.119 Station identification. 131
§97.121 Restricted operation. 132
Subpart C—Special Operations 132
§97.201 Auxiliary station. 132
§97.203 Beacon station. 132
§97.205 Repeater station. 133
§97.207 Space station. 133
§97.209 Earth station. 133
§97.211 Space telecommand station. 136
§97.213 Telecommand of an amateur station. 136
§97.215 Telecommand of model craft. 136
§97.217 Telemetry. 136
§97.219 Message forwarding system. 136
§97.221 Automatically controlled digital station. 137
Subpart D—Technical Standards 137
§97.301 Authorized frequency bands. 137
§97.303 Frequency sharing requirements. 141
§97.305 Authorized emission types. 144
§97.307 Emission standards. 146
§97.309 RTTY and data emission codes. 148
§97.311 SS emission types. 148
§97.313 Transmitter power standards. 148

§97.315 Certification of external RF power amplifiers..149
§97.317 Standards for certification of external RF power amplifiers. 149

Subpart E—Providing Emergency Communications...150
§97.401 Operation during a disaster...150
§97.403 Safety of life and protection of property. 150
§97.405 Station in distress. 150
§97.407 Radio amateur civil emergency service..150

Subpart F—Qualifying Examination Systems 151
§97.501 Qualifying for an amateur operator license. 151
§97.503 Element standards...152
§97.505 Element credit..152
§97.507 Preparing an examination..152
§97.509 Administering VE requirements. 152
§97.511 Examinee conduct. 153
§97.513 VE session manager requirements. 153
§§97.515 - 97.517 [Reserved] 153
§97.519 Coordinating examination sessions. 153
§97.521 VEC qualifications. 154
§97.523 Question pools...154
§97.525 Accrediting VEs...154
§97.527 Reimbursement for expenses. 154
Appendix 1 to Part 97—Places Where the Amateur Service is Regulated by the FCC...................154
Appendix 2 to Part 97—VEC Regions..155
§2.106, footnote US270..155

How to Renew Your License *156*

How to Apply for a Vanity Call Sign..*157*
Obtaining Vanity Call Signs 157
Call Sign Availability 157
General Rules 158

Amateur Radio Web Links *158*

Amateur Radio Glossary..*167*

Index *205*

Sample Station Log Page *210*

Azimuthal Map...*211*

By the Same Author..*212*

The Radio Amateur's Code

The Radio Amateur is

CONSIDERATE. He/She never knowingly operates in such a way as to lessen the pleasure of others.

LOYAL. He/She offers loyalty, encouragement and support to other amateurs, local clubs, the IARU Radio Society in his/her country, through which Amateur Radio in his/her country is represented nationally and internationally.

PROGRESSIVE. He/She keeps his/her station up to date. It is well-built and efficient. His/Her operating practice is above reproach.

FRIENDLY. He/She operates slowly and patiently when requested; offers friendly advice and counsel to beginners; kind assistance, cooperation and consideration for the interests of others. These are the marks of the amateur spirit.

BALANCED. Radio is a hobby, never interfering with duties owed to family, job, school or community.

PATRIOTIC. His/Her station and skills are always ready for service to country and community.

- adapted from the original Amateur's Code, written by Paul M. Segal, W9EEA, in 1928

Your License Privileges

Technician

Band	Frequencies	Modes
80 meters	3.525 – 3.600	CW
40 meters	7.025 – 7.125	CW
15 meters	21.025 – 21.200	CW
10 meters	28.000 – 28.300 28.300 – 28.500	CW, RTTY/data, 200 watts PEP maximum power CW, SSB phone, 200 watts PEP maximum power
Above 50 MHz	All amateur privileges	All legal modes
May not serve as a Volunteer Examiner		

General

Band	Frequencies	Modes
160, 60, 30 meters	All amateur privileges.	All legal modes.
80 meters	3.525 – 3.600	CW, RTTY, data
	3.800 – 4.000	CW, phone, image
40 meters	7.025 – 7.125	CW, RTTY, data
	7.175 – 7.300	CW, phone, image
20 meters	14.025 – 14.150	CW, RTTY, data
	14.225 – 14.350	CW, phone, image
15 meters	21.025 – 21.200	CW, RTTY, data
	21.275 – 21.450	CW, phone, image
17, 12, 10 meters	All amateur privileges.	All legal modes.
Above 50 MHz	All amateur privileges.	All legal modes.
As a Volunteer Examiner, may administer exams to Technician class candidates.		

Advanced

Band	Frequencies	Modes
160, 60, 30 meters	All amateur privileges.	All legal modes.
80 meters	3.525 – 3.600	CW, RTTY, data
	3.700 – 4.000	CW, phone, image
40 meters	7.025 – 7.125	CW, RTTY, data
	7.125 – 7.300	CW, phone, image
20 meters	14.025 – 14.150	CW, RTTY, data
	14.175 – 14.350	CW, phone, image
15 meters	21.025 – 21.200	CW, RTTY, data
	21.225 – 21.450	CW, phone, image
17, 12, 10 meters	All amateur privileges.	All legal modes.
Above 50 MHz	All amateur privileges.	All legal modes.
As a Volunteer Examiner, may administer exams to Technician Class and General Class candidates.		

Amateur Extra

Band	Frequencies	Modes
All amateur bands	All amateur privileges.	All legal modes.
As a Volunteer Examiner, may administer exams to candidates for all classes of license.		

Frequency Allocations & Band Plans

These tables consolidate information in the official ARRL band plans and *The Considerate Operator's Frequency Guide* published by the ARRL. There are minor discrepancies between those documents, and in those cases these tables show the official band plan information.

While the FCC assigned frequency allocations and modes are the law and must be followed, the divisions of those allocations known as the band plans are not allocations – they do not carry the force of law. From time to time, a high level of activity may result in stations operating outside these frequency ranges. Part 97 requires that we "must cooperate in selecting transmitting channels and in making the most effective use of the amateur service frequencies," and the band plans represent an effort in that direction.

- ■ Phone & Image or All Modes
- ▦ RTTY & Data Only
- ≋ CW Only
- ▥ Digipeaters Only
- ☐ SSB Phone Only

N = Novice Class License   G = General Class License
T = Technician Class License   X = Amateur Extra Class License

CW operation is permitted on any amateur frequency, provided the operator possesses the proper license for that frequency. MCW is permitted above 50.1 MHz, except for 144.0 MHz to 144.1 MHz and 219.0 MHz to 220.0 MHz. Test transmissions are authorized above 51 MHz, except for 219.0 MHz to 220.0 MHz.

Where, in adjacent ITU Regions or sub-Regions, a band of frequencies is allocated to different services of the same category (i.e., primary or secondary services), the basic principle is the equality of right to operate. Accordingly, stations of each service in one Region or sub-Region must operate so as not to cause harmful interference to any service of the same or higher category in the other Regions or sub-Regions.

Super Low Frequency (SLF)

2200 Meters 135.7 - 137.8 kHz

2200 Meters
RTTY, Data, Phone & Image — G, A, X
135.7 kHz — 137.8 kHz

General, Advanced, and Amateur Extra class:

135.7 - 137.8 kHz: CW, Phone, Image, RTTY/Data

1 watt EIRP (Equivalent Isotropically Radiated Power) maximum. Antennas used to transmit in the 2200 m band must not exceed 60 meters in height above ground level. (No ARRL band plan at this time. See below for IARU Region 1 plan.)

IARU 2200-Meter Band Plan	
135.7–136.0 kHz	Station Tests and transatlantic reception window
136.0–137.4 kHz	Telegraphy
137.4–137.6 kHz	Non-Telegraphy digital modes
137.6–137.8 kHz	Very slow telegraphy centered on 137.7 kHz

Ultra-Low Frequency (ULF)

630 Meters 472 kHz – 479 kHz

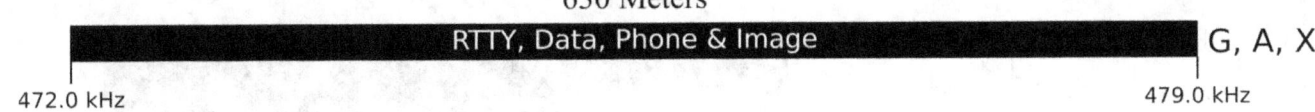

General, Advanced, and Amateur Extra class:

472 - 479 kHz: CW, Phone, Image, RTTY/Data

5 watts EIRP maximum, except in Alaska within 496 miles of Russia where the power limit is 1 watt EIRP. Antennas used to transmit in the 630 m band must not exceed 60 meters in height above ground level. (No band plan at this time.)

To operate on 2200 or 630 meters, amateurs must first register with the Utilities Technology Council online at https://utc.org/plc-database-amateur-notification-process/. You need only register once for each band.

Medium Frequency (MF)

160 Meters 1.8 – 2.0 MHz

General, Advanced, Amateur Extra licensees:

1.800 - 2.000 MHz: CW, Phone, Image, RTTY/Data

160-meter Band Plan	
1.800 - 2.000	CW
1.800 - 1.810	Digital Modes
1.810	CW QRP calling frequency
1.843 - 2.000	SSB, SSTV and other wideband modes
1.910	SSB QRP
1.995 - 2.000	Experimental
1.999 - 2.000	Beacons

Note: Amateurs in ITU Region 1 have privileges on the 160-meter band only from 1.810 MHz to 1.850 MHz.

High Frequency

80 Meters 3.5 – 4.0 MHz

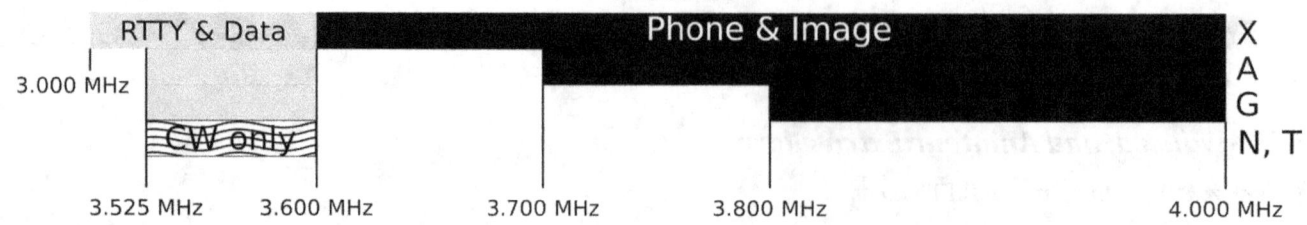

Novice and Technician classes:
3.525 - 3.600 MHz: CW Only

General class:
3.525 - 3.600 MHz: CW, RTTY/Data
3.800 - 4.000 MHz: CW, Phone, Image

Advanced class:
3.525 - 3.600 MHz: CW, RTTY/Data
3.700 - 4.000 MHz: CW, Phone, Image

Amateur Extra class:
3.500 - 3.600 MHz: CW, RTTY/Data
3.600 - 4.000 MHz: CW, Phone, Image

80-meter Band Plan	
3.500 – 3.510	CW DX window
3.560	QRP CW calling frequency
3.570 – 3.600	RTTY/Data
3.585 – 3.600	Automatically controlled data stations
3.590	RTTY/Data DX
3.570 – 3.600	RTTY/Data
3.790 – 3.800	DX window
3.845	SSTV
3.885	AM calling frequency
3.985	QRP SSB calling frequency

Note: In ITU Region 1, the amateur 80-meter band excludes frequencies above 3.800 MHz. In ITU Region 3, the amateur 80-meter band excludes frequencies above 3.900 MHz.

60 Meters 5.3 MHz, Channelized

Amateur radio has secondary access on USB only to five discrete 2.8-kHz-wide channels. Amateurs may not cause inference to and must accept interference from the Primary Government users, Hams planning to operate on 60 meters "must assure that their signal is transmitted on the channel center frequency." This means that amateurs should set their carrier frequency 1.5 kHz *lower* than the channel center frequency for USB transmissions.

General, Advanced and Amateur Extra classes:

Five channels, centered at 5332 kHz, 5348 kHz, 5358.5 kHz, 5373 kHz, and 5405 kHz.

Channel Center	USB Phone & Data Tuning Frequency	CW Tuning Frequency
5332 kHz	5330.5 kHz	5332 kHz
5348 kHz	5346.5 kHz	5348 kHz
5358.5 kHz	5357.0 kHz	5358.5 kHz
5373 kHz	5371.5 kHz	5373 kHz
5405 kHz (common US/UK)	5403.5 kHz	5405 kHz

Amateurs are permitted to use digital modes on 60 meters that comply with emission designator 60H0J2B (60 Hz bandwidth, phase shift keying.) This includes PSK31 as well as any RTTY signal with a bandwidth of less than 60 Hz.

They may also use modes that comply with emission designator 2K80J2D (2.8 kHz bandwidth, single sideband data), which includes any digital mode with a bandwidth of 2.8 kHz or less whose technical characteristics have been documented publicly, per Part 97.309(4) of the FCC Rules. Such modes would include PACTOR I, II or III, 300-baud packet, MFSK16, MT63, Contestia, Olivia, DominoEX and others.

Amateurs are allowed a maximum effective radiated power (ERP) of *100 W* on 60 meters. Radiated power must not exceed the equivalent of 100 W PEP transmitter output power into an antenna with a gain of 0 dBd.

* Only one signal at a time is permitted on any channel

* Maximum ERP -- Effective Radiated Power -- output is 100 W PEP

60-meter Band Plan	
5330.5 kHz	USB phone[1]
5332 kHz	CW/RTTY/data[2]
5346.5 kHz	USB phone[1]
5348 kHz	CW/RTTY/data[2]
5357.0 kHz	USB phone[1]
5358.5 kHz	CW/RTTY/data[2]
5371.5 kHz	USB phone[1]
5373 kHz	CW/RTTY/data[2]
5403.5 kHz	USB phone[1]
5405 kHz	CW/RTTY/data[2]

[1] USB is limited to 2.8 kHz bandwidth.

[2] CW and digital emissions must be centered 1.5 kHz above the channel frequencies indicated in the above chart.

40 Meters 7.0 MHz – 7.3 MHz

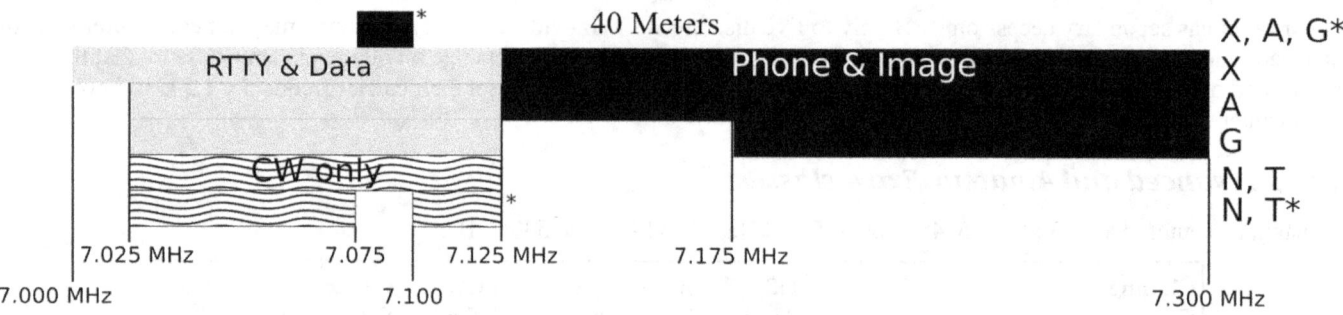

Novice and Technician classes:

7.025 - 7.125 MHz: CW only

General class:

7.025 - 7.125 MHz: CW, RTTY/Data
7.175 - 7.300 MHz: CW, Phone, Image

Advanced class:

7.025 - 7.125 MHz: CW, RTTY/Data
7.125 - 7.300 MHz: CW, Phone, Image

Amateur Extra class:

7.000 - 7.125 MHz: CW, RTTY/Data
7.125 - 7.300 MHz: CW, Phone, Image

40-meter Band Plan	
7.030	QRP CW calling frequency
7.040	RTTY/Data DX
7.070 - 7.125	RTTY/Data
7.100 – 7.105	Automatically controlled data stations
7.171	SSTV
7.173	D-SSTV
7.285	QRP SSB calling frequency
7.290	AM calling frequency

* Phone and Image modes are permitted between 7.075 and 7.100 MHz for FCC licensed stations in ITU Regions 1 and 3 and by FCC licensed stations in ITU Region 2 West of 130 degrees West longitude or South of 20 degrees North latitude. See Sections 97.305(c) and 97.307(f)(11). Novice and Technician licensees outside ITU Region 2 may use CW only between 7.025 and 7.075 MHz and between 7.100 and 7.125 MHz. 7.200 to 7.300 MHz is not available outside ITU Region 2. See Section 97.301(e). **These exemptions do not apply to any stations in the continental US.**

Note: International shortwave broadcasters in ITU Regions 1 and 2 are licensed to operate from 7.200 MHz to 7.300 MHz (and beyond.) The amateur 40-meter band in those regions does not include 7.200 to 7.300 MHz.

30 Meters 10.100 – 10.150 MHz

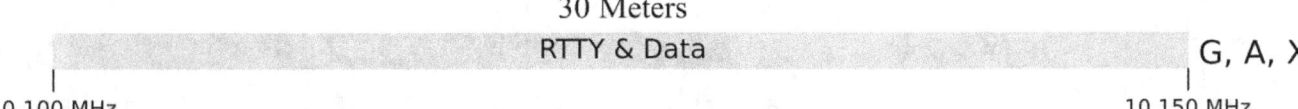

General, Advanced, Amateur Extra classes:

10.100 - 10.150 MHz: CW, RTTY/Data

30-meter Band Plan	
10.130 - 10.140	CW, RTTY
10.140 - 10.150	CW, Packet

Maximum power on 30 meters is 200 watts PEP. Amateurs must avoid interference to the fixed service outside the US.

20 Meters 14.000 – 14.350 MHz

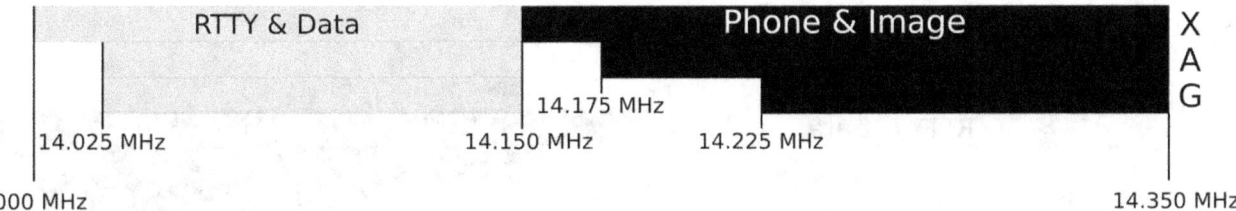

General class:

14.025 - 14.150 MHz: CW, RTTY/Data
14.225 - 14.350 MHz: CW, Phone, Image

Amateur Extra class:

14.000 - 14.150 MHz: CW, RTTY/Data
14.150 - 14.350 MHz: CW, Phone, Image

Advanced class:

14.025 - 14.150 MHz: CW, RTTY/Data
14.175 - 14.350 MHz: CW, Phone, Image

20-meter Band Plan	
14.060	QRP CW calling frequency
14.070 - 14.095	RTTY/Data
14.095 - 14.0995	Packet – automatically controlled data stations
14.100	**IBP/NCDXF Beacons**
14.1005 - 14.112	Packet – automatically controlled data stations
14.230	SSTV
14.233	D-SSTV
14.236	Digital Voice
14.285	QRP SSB calling frequency
14.286	AM calling frequency

17 Meters 18.068 – 18.168 MHz

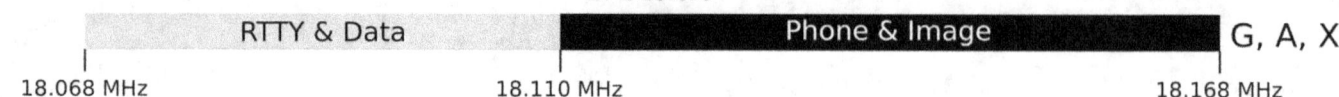

General, Advanced, Amateur Extra classes:

18.068 - 18.110 MHz: CW, RTTY/Data
18.110 - 18.168 MHz: CW, Phone, Image

17-meter Band Plan	
18.100 - 18.105	RTTY/Data
18.105 - 18.110	Packet – automatically controlled data stations
18.110	**IBP/NCDXF Beacons**
18.162.5	Digital Voice

15 Meters 21.000 – 21.450 MHz

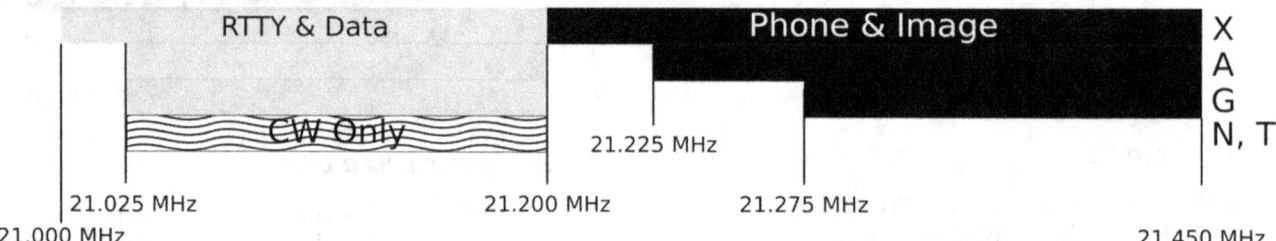

Novice and Technician classes:

21.025 - 21.200 MHz: CW Only

General class:

21.025 - 21.200 MHz: CW, RTTY/Data
21.275 - 21.450 MHz: CW, Phone, Image

Advanced class:

21.025 - 21.200 MHz: CW, RTTY/Data
21.225 - 21.450 MHz: CW, Phone, Image

Amateur Extra class:

21.000 - 21.200 MHz: CW, RTTY/Data
21.200 - 21.450 MHz: CW, Phone, Image

15-meter Band Plan	
21.060	QRP CW calling frequency
21.070 - 21.110	RTTY/Data
21.090 – 21.100	Automatically controlled data stations
21.150	**IBP/NCDXF Beacons**
21.340	SSTV
21.385	QRP SSB calling frequency

12 Meters 24.890 – 24.990 MHz

General, Advanced, Amateur Extra classes:

24.890 - 24.930 MHz: CW, RTTY/Data
24.930 - 24.990 MHz: CW, Phone, Image

12-meter Band Plan	
24.920 - 24.925	RTTY/Data
24.925 - 24.930	Packet
24.930	**IBP/NCDXF Beacons**

10 Meters 28.000 – 29.700 MHz

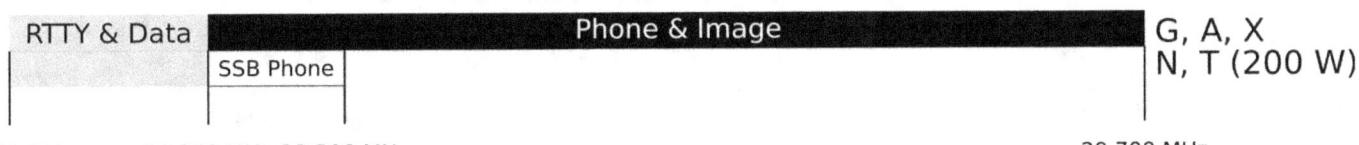

Novice and Technician classes:

28.000 - 28.300 MHz: CW, RTTY/Data--Maximum power 200 watts PEP
28.300 - 28.500 MHz: CW, Phone--Maximum power 200 watts PEP

General, Advanced, Amateur Extra classes:

28.000 - 28.300 MHz: CW, RTTY/Data
28.300 - 29.700 MHz: CW, Phone, Image

10-meter Band Plan			
28.000 - 28.070	CW	28.300 - 29.300	Phone
28.060	QRP CW calling frequency	28.680	SSTV
28.070 - 21.120	RTTY/Data	29.000 - 29.200	AM
28.120 – 28.189	Automatically controlled data stations	29.300 - 29.510	Satellite Downlinks
28.150 - 28.190	CW	29.520 - 29.580	Repeater Inputs
28.190 – 28.225	Beacons	29.600	FM Simplex
28.200	**IBP/NCDXF Beacons**	29.620 - 29.680	Repeater Outputs

Unofficial standard 10-meter repeater offset is –100 kHz.

Very High Frequencies (VHF)

6 Meters 50.000 – 54.000 MHz

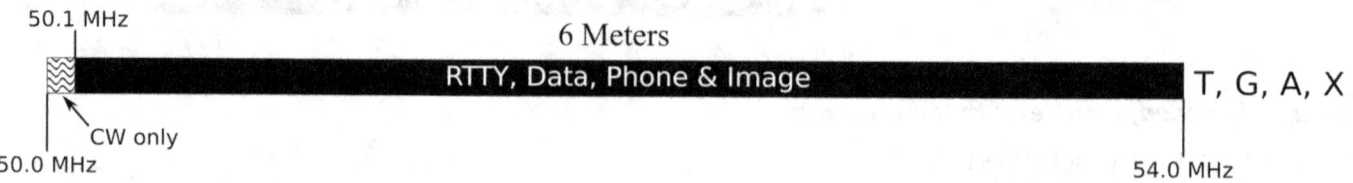

All Amateurs except Novice class:

50.0 - 50.1 MHz: CW Only
50.1 - 54.0 MHz: CW, Phone, Image, MCW, RTTY/Data

6-meter Band Plan				
50.0 - 50.1	CW, beacons	52.0 - 52.48	Repeater inputs (except as noted; 23 channels)
50.06 - 50.08	Beacon sub-band	52.02, 52.04	FM simplex
50.1 - 50.3	SSB, CW	52.2	Test pair (input)
50.10 - 50.125	DX window	52.5 - 52.98	Repeater output (except as noted; 23 channels)
50.125	SSB calling	52.525	Primary FM simplex
50.3 - 50.6	All modes	52.54	Secondary FM simplex
50.6 - 50.8	Nonvoice communications	52.7	Test pair (output)
50.62	Digital (packet) calling	53.0 - 53.48	Repeater inputs (except as noted; 19 channels)
50.8 - 51.0	Radio remote control (20 -kHz channels)	53.0	Remote base FM simplex
51.0 - 51.1	Pacific DX window	53.02	Simplex
51.12 - 51.48	Repeater inputs (19 channels)	53.1, 53.2, 53.3, 53.4	Radio remote control
51.12 - 51.18	Digital repeater inputs	53.5 - 53.98	Repeater outputs (except as noted; 19 channels)
51.5 - 51.6	Simplex (six channels)	53.5, 53.6, 53.7, 53.8	Radio remote control
51.62 - 51.98	Repeater outputs (19 channels)	53.52, 53.9	Simplex
51.62 - 51.68	Digital repeater outputs		

6-meter repeater offsets vary, but the most common are -0.5 MHz, and -1.0 MHz.

Note: Amateurs in ITU Region 1 do not have privileges on any portion of the 6-meter band.

2 Meters 144.000 – 148.000 MHz

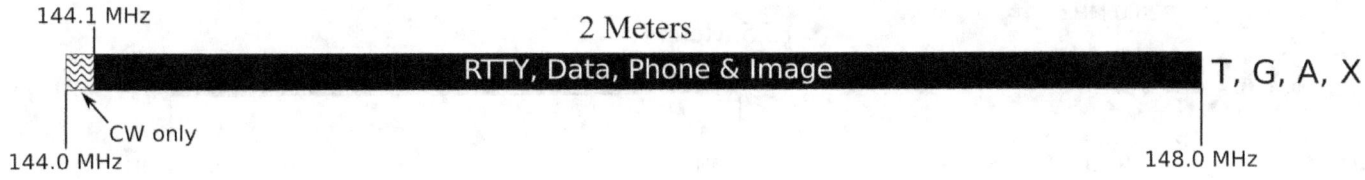

All Amateurs except Novice class

144.0 - 144.1 MHz: CW Only
144.1 - 148.0 MHz: CW, Phone, Image, MCW, RTTY/Data

2-meter Band Plan			
144.00 - 144.05	EME (CW)	145.20 - 145.50	FM repeater outputs
144.05 - 144.10	General CW and weak signals	145.50 - 145.80	Miscellaneous and experimental modes
144.10 - 144.20	EME and weak-signal SSB	145.80 - 146.00	OSCAR subband
144.200	National calling frequency	146.01 - 146.37	Repeater inputs
144.200 - 144.275	General SSB operation	146.40 - 146.58	Simplex
144.275 - 144.300	Propagation beacons	146.52	National Simplex Calling Frequency
144.30 - 144.50	OSCAR subband	146.61 - 146.97	Repeater outputs
144.50 - 144.60	Linear translator inputs	147.00 - 147.39	Repeater outputs
144.60 - 144.90	FM repeater inputs	147.42 - 147.57	Simplex
144.90 - 145.10	Weak signal and FM simplex (145.01, 03, 05, 07, 09 are widely used for packet)	147.60 - 147.99	Repeater inputs
145.10 - 145.20	Linear translator outputs		

The frequency 146.40 MHz is used in some areas as a repeater input. This band plan has been proposed by the ARRL VHF-UHF Advisory Committee.

Band plans in major metropolitan areas may differ from this plan.

Standard repeater offsets for the 2-meter band are 600 kHz. For repeater outputs below 147.000 MHz, -600 kHz, for those above 147.000 MHz, +600 kHz. Your local offsets may vary.

Note: In ITU Region 1, the 2-meter band excludes frequencies from 146.000 MHz to 148 MHz.

1.25 Meters 219.000 – 225.000 MHz

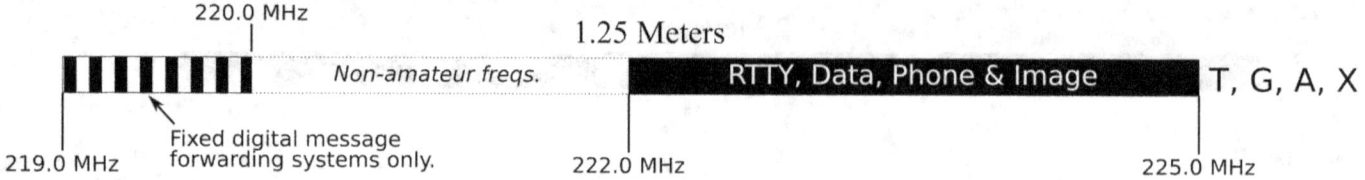

All Amateurs except Novice class

219.0 – 220.0 MHz

The FCC has allocated 219 - 220 MHz to amateur use on a secondary basis. This allocation is *only* for fixed digital message forwarding systems operated by all licensees **except Novice class**.

Amateur operations must not cause interference to, and must accept interference from, primary services in this and adjacent bands.

Amateur stations are limited to 50 W PEP output and 100 kHz bandwidth.

Automated Maritime Telecommunications Systems (AMTS) stations are the primary occupants in this band. Amateur stations within 398 miles of an AMTS station must notify the station in writing at least 30 days prior to beginning operations. Amateur stations within 50 miles of an AMTS station must get permission in writing from the AMTS station before beginning operations.

The FCC requires that amateur operators provide written notification including the station's geographic location to the ARRL for inclusion in a database at least 30 days before beginning operations. See Section 97.303(e) of the FCC Rules.

Novice, Technician, General, Advanced, Amateur Extra classes:

222.00 - 225.00 MHz: CW, Phone, Image, MCW, RTTY/Data

Novices are limited to 25 watts PEP output

1.25 Meter Band Plan	
222.0 - 222.150	Weak-signal modes
222.0 - 222.025	EME
222.05 - 222.06	Propagation beacons
222.1	SSB & CW calling frequency
222.10 - 222.15	Weak-signal CW & SSB
222.15 - 222.25	Local coordinator's option; weak signal, ACSB, repeater inputs, control
222.25 - 223.38	FM repeater inputs only
223.40 - 223.52	FM simplex
223.52 - 223.64	Digital, packet
223.64 - 223.70	Links, control
223.71 - 223.85	Local coordinator's option; FM simplex, packet, repeater outputs
223.85 - 224.98	Repeater outputs only

1.25-meter repeater offsets are almost universally -1.6 MHz.

Note: Amateurs in ITU Regions 1 and 3 do not have any privileges in the 1.25 meter band.

Ultra High Frequencies (UHF)

70 Centimeters 420.000 – 450.000 MHz

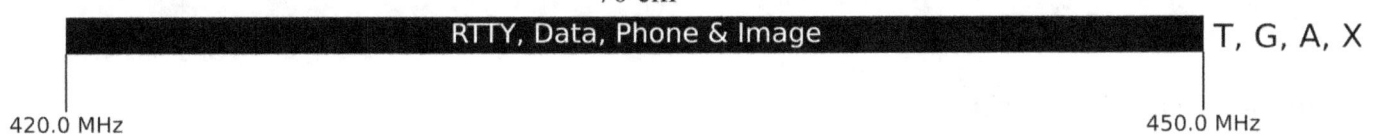

All Amateurs except Novice class:

420.0 - 450.0 MHz: CW, Phone, Image, MCW, RTTY/Data

70 cm Band Plan	
420.00 - 426.00	ATV repeater or simplex with 421.25 MHz video carrier control links and experimental
426.00 - 432.00	ATV simplex with 427.250 MHz video carrier frequency
432.00 - 432.07	EME (Earth-Moon-Earth)
432.07 - 432.10	Weak-signal CW
432.10	70-cm calling frequency
432.10 - 432.30	Mixed-mode and weak-signal work
432.30 - 432.40	Propagation beacons
432.40 - 433.00	Mixed-mode and weak-signal work
433.00 - 435.00	Auxiliary/repeater links
435.00 - 438.00	Satellite only (internationally)
438.00 - 444.00	ATV repeater input with 439.250 MHz video carrier frequency and repeater links
442.00 - 445.00	Repeater inputs and outputs (local option)
445.00 - 447.00	Shared by auxiliary and control links, repeaters and simplex (local option)
446.00	National simplex frequency
447.00 - 450.00	Repeater inputs and outputs (local option)

70 cm repeater offsets are less consistent nationwide than 2 m repeater offsets. As a very general rule, offsets are 5 MHz. For repeater outputs below 445.000 MHz, +5 MHz, for those above 445.000 MHz, -5 MHz. Your local offsets may vary.

Amateur stations may not transmit on the 420 – 430 MHz frequency segments if they are located north of Line A, approximately 50 miles south of the Canadian border. Line "A" and Line "C" check is here: http://wireless.fcc.gov/uls/index.htm?job=line_a_c

Amateur stations transmitting in the 430 - 450 MHz segment must not cause harmful interference to, and must accept interference from, stations authorized by other nations in the radiolocation service.

Note: In ITU Regions 1 and 3, the 70 cm band is limited to 430.000 MHz to 440.000 MHz.

33 Centimeters

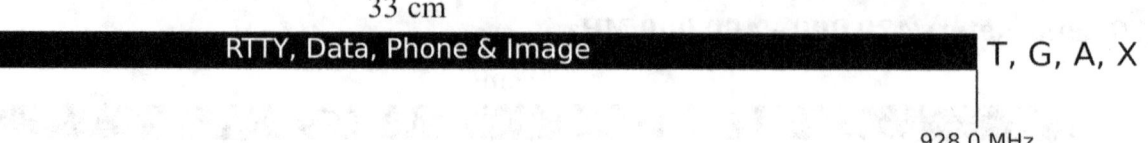

All Amateurs except Novice class:

902.0 - 928.0 MHz: CW, Phone, Image, MCW, RTTY/Data

33 cm Band Plan				
Frequency	Mode	Functional Use	Comments
902.000 - 902.075	FM /other including DV or CW/SSB	Repeater inputs 25 MHz split paired with those in 927.000 - 927.075 or Weak signal[1]	12.5 kHz channel spacing
902.075 - 902.100	CW/SSB	Weak signal	
902.100	CW/SSB	Weak signal calling	Regional option
902.100 - 902.125	CW/SSB	Weak signal	
902.125 - 903.000	FM/other including DV	Repeater inputs 25 MHz split paired with those in 927.1250 - 928.0000	12.5 kHz channel spacing
903.000 - 903.100	CW/SSB	Beacons and weak signal	
903.100	CW/SSB	Weak signal calling	Regional option
903.100 - 903.400	CW/SSB	Weak signal	
903.400 - 909.000	Mixed modes	Mixed operations including control links	
909.000 - 915.000	Analog/digital	Broadband multimedia including ATV, DATV and SS[2, 3]	
915.000 - 921.000	Analog/digital	Broadband multimedia including ATV, DATV and SS[2, 3]	
921.000 - 927.000	Analog/digital	Broadband multimedia including ATV, DATV and SS[2, 3]	
927.000 - 927.075	FM /other including DV	Repeater outputs 25 MHz split paired with those in 902.0000 - 902.0750	12.5 kHz channel spacing
927.075 - 927.125	FM/other including DV	Simplex	
927.125 - 928.000	FM/other including DV	Repeater outputs 25 MHz split paired with those in 902.125 - 903.000[4, 5]	12.5 kHz channel spacing

Significant regional variations in both current band utilization and the intensity and frequency distribution of noise sources preclude one plan that is suitable for all parts of the country. These variations will require many regional frequency coordinators to maintain band plans that differ in some respects from any national plan.

As with all band plans, locally coordinated plans always take precedence over any general recommendations such as a national band plan.

1) May be used for either repeater inputs or weak-signal as regional needs dictate.

2) Division into channels and/or separation of uses within these segments may be done regionally based on needs and usage, such as for 2 MHz-wide digital TV.

3) These segments may also be designated regionally to accommodate alternative repeater splits.

4) Simplex FM calling frequency 927.500 or regionally selected alternative.

5) Additional FM simplex frequencies may be designated regionally.

33 cm repeater offsets are almost all -25 MHz.

Note: Amateurs in ITU Regions 1 and 3 have no privileges in the 33 cm band.

23 Centimeters 1240.000 – 1300.000 MHz

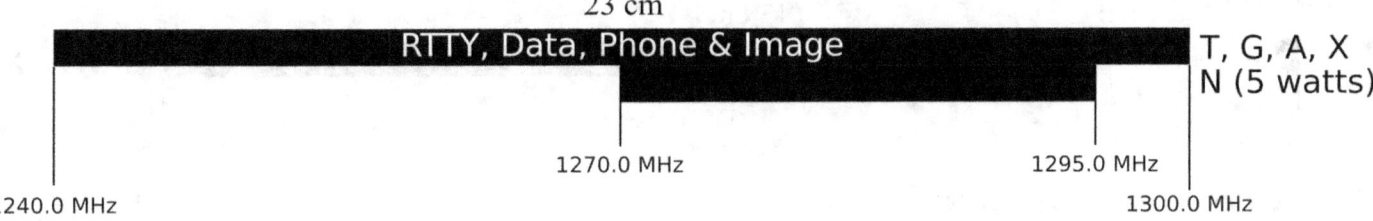

Novice class:

1270 - 1295 MHz: CW, phone, Image, MCW, RTTY/Data (maximum power, 5 watts PEP)

All Amateurs except Novice class:

1240 - 1300 MHz: CW, Phone, Image, MCW, RTTY/Data

23 cm Band Plan		
Frequency Range	Suggested Emission Types	Functional Use
1240.00 - 1246.000	ATV	ATV Channel #1
1246.000 - 1248.000	FM, digital	Point-to-point links paired with 1258.000 - 1260.000
1248.000 - 1252.000	Digital	
1252.000 - 1258.000	ATV	Amateur TV Channel #2
1258.000 - 1260.000	FM, digital	Point-to-point links paired with 1246.000 - 1248.000
1240.000 - 1260.000	FM ATV	Regional option
1260.000 - 1270.000	Various	Satellite uplinks, Experimental, Simplex ATV
1270.000 - 1276.000	FM, digital	Repeater inputs, 25 kHz channel spacing, paired with 1282.000 - 1288.000
1270.000 - 1274.000	FM, digital	Repeater inputs, 25 kHz channel spacing, paired with 1290.000 - 1294.000 (Regional
1276.000 - 1282.000	ATV	ATV Channel #3
1282.000 - 1288.000	FM, digital	Repeater outputs, 25 kHz channel spacing, paired with 1270.000 - 1276.000
1288.000 - 1294.000	Various	Broadband Experimental, Simplex ATV
1290.000 - 1294.000	FM, digital	Repeater outputs, 25 kHz channel spacing, paired with 1270.000 - 1274.000 (Regional
1294.000 - 1295.000	FM	FM simplex
	FM	National FM simplex calling frequency 1294.500
1295.000 - 1297.000		Narrow Band Segment
1295.000 - 1295.800	Various	Narrow Band Image, Experimental
1295.800 - 1296.080	CW, SSB, digital	EME
1296.080 - 1296.200	CW, SSB	Weak Signal
1296.100	CW, SSB	CW, SSB calling frequency
1296.200 - 1296.400	CW, digital	Beacons
1296.400 - 1297.000	Various	General Narrow Band
1297.000 - 1300.000	Digital	

The need to avoid harmful interference to FAA radars may limit amateur use of certain frequencies in the vicinity of the radars.

The most common 23 cm repeater offset is -12 MHz.

13 Centimeters 2300.000 – 2450.000 MHz

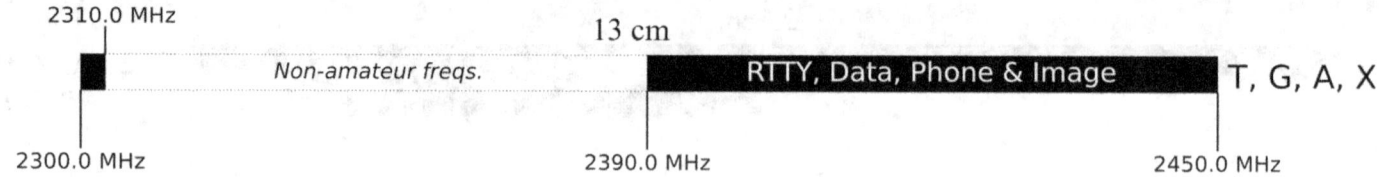

All amateurs except Novice class

2300 – 2310 MHz, all modes

2390 – 2450 MHz, all modes

13 cm Band Plan		
Frequency Range	Emission Bandwidth	Functional Use
2300.000 - 2303.000	0.05 - 1.0 MHz	Analog & Digital, including full duplex; paired with 2390 - 2393
2303.000 - 2303.750	< 50 kHz	Analog & Digital; paired with 2393 - 2393.750
2303.75 - 2304.000		SSB, CW, digital weak-signal
2304.000 - 2304.100	3 kHz or less	Weak Signal EME Band
2304.10 - 2304.300	3 kHz or less	SSB, CW, digital weak-signal (Note 1)
2304.300 - 2304.400	3 kHz or less	Beacons
2304.400 - 2304.750	6 kHz or less	SSB, CW, digital weak-signal & NBFM
2304.750 - 2305.000	< 50 kHz	Analog & Digital; paired with 2394.750 - 2395
2305.000 - 2310.000	0.05 - 1.0 MHz	Analog & Digital, paired with 2395 - 2400 (Note 2)
2310.000 - 2390.000		**NON-AMATEUR**
2390.000 - 2393.000	0.05 - 1.0 MHz	Analog & Digital, including full duplex; paired with 2300 - 2303
2393.000 - 2393.750	< 50 kHz	Analog & Digital; paired with 2303 - 2303.750
2393.750 - 2394.750		Experimental
2394.750 - 2395.000	< 50 kHz	Analog & Digital; paired with 2304.750 - 2305
2395.000 - 2400.000	0.05 - 1.0 MHz	Analog & Digital, including full duplex; paired with 2305 - 2310
2400.000 - 2410.000	6 kHz or less	Amateur Satellite Communications
2410.000 - 2450.000	22 MHz max.	Broadband Modes (Notes 3, 4)

1) 2304.100 is the National Weak-Signal Calling Frequency

2) 2305 - 2310 is allocated on a primary basis to Wireless Communications Services (Part 27). Amateur operations in this segment, which are secondary, may not be possible in all areas.

3) Broadband segment may be used for any combination of high-speed data (e.g. 802.11 protocols), Amateur Television and other high-bandwidth activities. Division into channels and/or separation of uses within this segment may be done regionally based on needs and usage.

4) 2424.100 is the Japanese EME transmit frequency

Super High Frequency (SHF)

9 Centimeters 3300.000 – 3500.000 MHz

9 cm
RTTY, Data, Phone & Image   T, G, A, X
3300.0 MHz — 3500.0 MHz

All amateurs except Novice class

3300 - 3500 MHz, all modes

9 cm Band Plan									
Level I - Major Band Divisions			Level II - Sub-Band Divisions			Level III	Suggested Emission Types[1]	Suggested Emission B.W.	Functional Use
Frequency Range (MHz)			*Frequency Range (MHz)*			*Specific Freq.*			
From	*To*	*Width*	*From*	*To*	*Width*	*MHz*			
3300.000	3309.000	9.0					Analog & Digital, including Full Duplex	0.1 - 1.0 MHz	Analog & Digital; paired with 3430.0 - 3439.0; 130 MHz Split
3309.000	3310.000	1.0							Experimental
3310.000	3330.000	20.0					Analog & Digital, including Full Duplex	> 1.0 MHz	Analog & Digital; paired with 3410.0 - 3430.0; 100 MHz Split
3330.000	3332.000	2.0							Experimental
3332.000	**3339.000**	**7.0**							**RADIO ASTRONOMY PROTECTED BAND (Note 4)**
3339.000	3345.800	6.8					Analog & Digital, including Full Duplex	0.1 - 1.0 MHz	Analog & Digital; paired with 3439.0 - 3445.8; 100 MHz Split
3345.800	**3352.500**	**6.7**	**RADIO ASTRONOMY PROTECTED BAND (Note 4)**							
3352.500	3355.000	2.5					Analog & Digital, including Full Duplex	0.05 - 0.2 MHz	Analog & Digital; paired with 3452.5 - 3455.0; 100 MHz Split
3355.000	3357.000	2.0							Experimental
3357.000	3360.000	3.0					Analog & Digital, including Full Duplex	50 kHz or less	Analog & Digital; paired with 3457.0 - 3460.0

9 cm Band Plan

Level I - Major Band Divisions			Level II - Sub-Band Divisions			Level III	Suggested Emission Types[1]	Suggested Emission B.W.	Functional Use
Frequency Range (MHz)			*Frequency Range (MHz)*			*Specific Freq.*			
From	*To*	*Width*	*From*	*To*	*Width*	*MHz*			
3360.000	3400.000	40.0					OFDM, others	22 MHz max.	Broadband Modes (Note 3)
			3360.000	3380.000	20.0		ATV		Amateur Television of all authorized modulation standards/formats at local option
3400.000	3410.000	10.0					CW, SSB, NBFM	6 kHz or less	Amateur Satellite Communications
			3400.000	3400.300	0.3		CW, SSB, Digital	3 kHz or less	Weak Signal EME Band
			3400.300	3401.000	0.7		CW, SSB, Digital	3 kHz or less	Terrestrial Weak Signal Band - Future (Note 2)
						3400.100	CW, SSB, Digital		EME Calling Frequency
3410.000	3430.000	20.0					Analog & Digital, including Full Duplex	> 1.0 MHz	Analog & Digital; paired with 3310.0 - 3330.0; 100 MHz Split
3430.000	3439.000	9.0					Analog & Digital, including Full Duplex	0.1 - 1.0 MHz	Analog & Digital; paired with 3300.0 - 3309.0; 130 MHz Split
3439.000	3445.800	6.8					Analog & Digital, including Full Duplex	0.1 - 1.0 MHz	Analog & Digital; paired with 3339.0 - 3345.8; 100 MHz Split
3445.800	3452.500	6.7							Experimental
3452.500	3455.000	2.5					Analog & Digital, including Full Duplex	0.05 - 0.2 MHz	Analog & Digital; paired with 3352.5 - 3355.0; 100 MHz Split
3455.000	3455.500	0.5						100 kHz or less	Crossband linear translator (input or output)

9 cm Band Plan

Level I - Major Band Divisions			Level II - Sub-Band Divisions			Level III	Suggested Emission Types[1]	Suggested Emission B.W.	Functional Use
Frequency Range (MHz)			*Frequency Range (MHz)*			*Specific Freq.*			
From	*To*	*Width*	*From*	*To*	*Width*	*MHz*			
3455.500	3457.000	1.5					CW, SSB, NBFM, Digital	6 kHz or less	Terrestrial Weak Signal Band - Legacy (Note 2)
						3456.100		6 kHz or less	Weak Signal Terrestrial Calling Frequency
			3456.300	3457.000	0.1		CW, Digital	1 kHz or less	Propagation Beacons
3457.000	3460.000	3.0					Analog & Digital, including Full Duplex	50 kHz or less	Analog & Digital; paired with 3357.0 - 3360.0; 100 MHz Split
3460.000	3500.000	40.0					OFDM, others	22 MHz max.	Broadband Modes (Note 3)
			3460.000	3480.000	20.0		ATV		Amateur Television of all authorized modulation standards/formats at local option

1) Includes all other emission modes authorized in the 9 cm amateur band whose necessary bandwidth does not exceed the suggested bandwidths listed.

2) Weak Signal Terrestrial legacy users are encouraged to move to 3400.3 to 3401.0 MHz as time and resources permit.

3) Broadband segments may be used for any combination of high-speed data (e.g. 802.11 protocols), Amateur Television and other high-bandwidth activities. Division into channels and/or separation of uses within these segments may be done regionally based on need and usage.

4) Per ITU RR 5.149 from WRC-07, these band segments are also used for Radio Astronomy. Amateur use of these frequencies should be first coordinated with the National Science Foundation (esm@nsf.gov).

Note: In ITU Region 1, amateurs have no privileges in the 9 cm band.

5 Centimeters 5650.000 – 5925.000 MHz

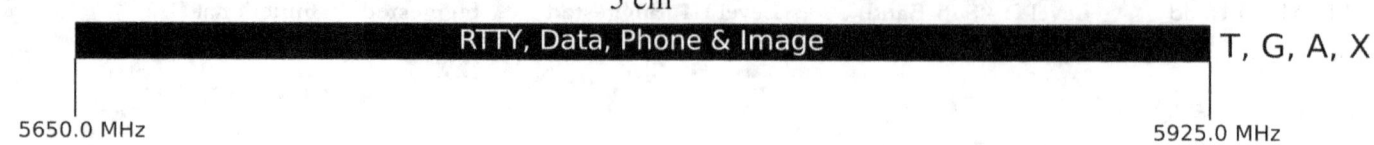

All amateurs except Novice class

5650.0 - 5925.0 MHz, all modes

5 cm Band Plan		
Frequency Range	Emission Bandwidth	Functional Use
5650.0 - 5670.0		Amateur Satellite; Up-Link Only
5650.0 - 5675.0	0.05 - 1.0 MHz	Experimental
5675.0 - 5750.0	≥ 1.0 MHz	Analog & Digital; paired with 5850 - 5925 MHz (Note 2)
5750.0 - 5756.0	≥ 25 kHz and < 1 MHz	Analog & Digital; paired with 5820 - 5826 MHz
5756.0 - 5759.0	≤ 50 kHz	Analog & Digital; paired with 5826 - 5829 MHz
5759.0 - 5760.0	< 6 kHz	SSB, CW, Digital Weak-Signal
5760.0 - 5760.1	< 3 kHz	EME
5760.1 - 5760.3	< 6 KHz	SSB, CW, Digital Weak-Signal (Note 1)
5760.3 - 5760.4	< 3 KHz	Beacons
5760.4 - 5761.0	< 6 KHz	SSB, CW, Digital Weak-Signal
5761.0 - 5775.0	≤ 50 kHz	Experimental
5775.0 - 5800.0	≥ 100 kHz	Experimental
5800.0 - 5820.0		Experimental
5820.0 - 5826.0	≥ 25 kHz and < 1 MHz	Analog & Digital; paired with 5750 - 5756 MHz
5826.0 - 5829.0	≤ 50 kHz	Analog & Digital; paired with 5756 - 5759 MHz
5829.0 - 5850.0	0.05 - 1.0 MHz	Experimental
5830.0 - 5850.0		Amateur Satellite; Down-Link Only
5850.0 - 5925.0	≥ 1.0 MHz	Analog & Digital; paired with 5675 - 5750 MHz (Note 2)

1) 5760.1 is the National Weak-Signal Calling Frequency

2) Broadband segment may be used for any combination of high-speed data (eg: 802.11 protocols), Amateur Television and other high-bandwidth activities. Division into channels and/or separation of uses within this segment may be done regionally based on needs and usage.

Note: In ITU Regions 1 and 3, the 5 cm band is limited to 5.650 GHz to 5.850 GHZ.

3 Centimeters 10000.000 – 10050.000 MHz

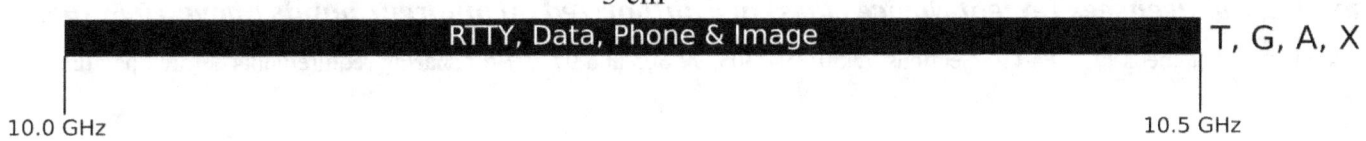

All amateurs except Novice class

10000.000 - 10500.000 MHz, all modes.

3 cm Band Plan		
Frequency Range	**Emission Bandwidth**	**Functional Use**
10000.00 - 10050.000		Experimental
10050.000 - 10100.000	≤ 100 kHz	Analog & Digital; paired with 10300 - 10350
10100.000 - 10115.000	≥ 25 kHz and < 1 MHz	Analog & Digital; paired with 10350 - 10365
10115.000 - 10117.000	≤ 50 kHz	Analog & Digital; paired with 10365 - 10367
10117.000 - 10120.000		Experimental
10120.000 - 10125.000	≤ 50 kHz	Analog & Digital; paired with 10370 - 10375
10125.000 - 10200.000	≥ 1 MHz	Analog & Digital; paired with 10375 - 10450 (Note 2)
10200.000 - 10300.000		Wideband Gunnplexers
10300.000 - 10350.000	≤ 100 kHz	Analog & Digital; paired with 10050 - 10100
10350.000 - 10365.000	≥ 25 kHz and < 1 MHz	Analog & Digital; paired with 10100 - 10115
10365.000 - 10367.000	≤ 50 kHz	Analog & Digital; paired with 10115 - 10117
10367.000 - 10368.300	6 kHz or less	SSB, CW, Digital Weak-Signal & NBFM (Note 1)
10368.300 - 10368.400	6 kHz or less	Beacons
10368.400 - 10370.000	6 kHz or less	SSB, CW, Digital Weak-Signal & NBFM
10370.000 - 10375.000	≤ 50 kHz	Analog & Digital; paired with 10120 - 10125
10375.000 - 10450.000	≥ 1 MHz	Analog & Digital; paired with 10125 - 10200 (Note 2)
10450.000 - 10500.000		Space, Earth & Telecommand Stations

1) 10368.100 is the National Weak-Signal Calling Frequency.

2) Broadband segment may be used for any combination of high-speed data (eg: 802.11 protocols), Amateur Television and other high-bandwidth activities. Division into channels and/or separation of uses within this segment may be done regionally based on needs and usage.

Above 10.50 GHz

All modes and licensees (except Novice class) are authorized on amateur bands above 10.5 GHz.

US amateurs must check FCC Part 97 Sections 97.301, 97.303, 97.305 and 97.307 for sharing requirements before operating.

24.0 - 24.25 GHz
47.0 - 47.2 GHz
76.0 - 81.0 GHz*
122.25 - 123.00 GHz
134 - 141 GHz
241 - 250 GHz
All above 300 GHz

*Amateur operation at 76 - 77 GHz has been suspended until the FCC can determine that interference will not be caused to vehicle radar systems.

Popular HF & 6-meter Digital Mode Frequencies

160 Meters	
Mode	Frequency (MHz)
FT4	-
FT8	1.84
JT65	1.838, 1.805 (alternate)
JT9	1.839
MFSK (OLIVIA)	1.80875, 1.80925, 1.84075, 1.84125 Calling frequency: 1.8254
PSK31	1.807, 1.838
SSTV	1.89

80 Meters	
Mode	Frequency (MHz)
FT4	3.575
FT8	3.573
JT65	3.570
JT9	3.572
MFSK (OLIVIA)	3577.75, 3583.25, 3522.75 (16/500 frequencies) 3578.0, 3616.0, 3523.0, 3621.0 (32/1000 frequencies) Calling frequency: 3.5714
PSK31	3.580
SSTV	3.845

60 Meters	
Mode	Frequency (MHz)
FT4	-
FT8	5.357
JT65	-
JT9	-
MFSK (OLIVIA)	5.3665
PSK31	-
SSTV	-

40 Meters	
Mode	Frequency (MHz)
FT4	7.0475
FT8	7.074
JT65	7.076
JT9	7.078
MFSK (OLIVIA)	7.02625, 7.04325, 7.07325, 7.07675 Calling frequency: 7.0714

40 Meters	
PSK31	7.040
SSTV	7.171

30 Meters	
Mode	Frequency (MHz)
FT4	10.140
FT8	10.136
JT65	10.138
JT9	10.140
MFSK (OLIVIA)	10.13925, 10.14225, 10.14325 Calling frequency: 10.1414
PSK31	10.141
SSTV	-

20 Meters	
Mode	Frequency (MHz)
FT4	14.080
FT8	14.074
JT65	14.076
JT9	14.078
MFSK (OLIVIA)	14.0764, 14.0754, 14.0784 (16/500 frequencies) 14.1065, 14.1075, 14.1085 (32/1000 frequencies) Calling frequency: 14.0714
PSK31	14.070
SSTV	14.230, 14.233

17 Meters	
Mode	Frequency (MHz)
FT4	18.104
FT8	18.100
JT65	18.102
JT9	18.104
MFSK (OLIVIA)	18.1034, 18.1044 Calling frequency: 18.1014
PSK31	18.098
SSTV	-

15 Meters	
Mode	Frequency (MHz)
FT4	21.140
FT8	21.074
JT65	21.076
JT9	21.078
MFSK (OLIVIA)	21.08725, 21.08775, 21.13025 (16/500 frequencies) 21.1535, 21.1545 (32/1000 frequencies) Calling frequency: 21.0714
PSK31	21.070
SSTV	21.340

12 Meters	
Mode	Frequency (MHz)
FT4	24.919
FT8	24.915
JT65	24.917
JT9	24.919
MFSK (OLIVIA)	24.92225 Calling frequency: 24.9214
PSK31	24.920
SSTV	-

10 Meters	
Mode	Frequency (MHz)
FT4	28.180
FT8	28.074
JT65	28.076
JT9	28.078
MFSK (OLIVIA)	28.07675, 28.07725 Calling frequency: 28.1214
PSK31	28.070, 28.120
SSTV	28.690, 28.700

6 Meters	
Mode	Frequency (MHz)
FT4	50.680
FT8	50.313
JT65	50.310
JT9	50.312
MFSK (OLIVIA)	50.08725, 50.28725, 50.29225
PSK31	50.305
SSTV	50.680

Family Radio Service and General Mobile Radio Service Frequencies

Note: For most ham radios in most circumstances, these are not – by the letter of the law – legal frequencies for us to use. In emergencies, however, we use whatever we need to use to prevent harm to life, limb or property.

Service	Frequency MHz	Channel #	FRS Max Power/Bandwidth	GMRS Max Power/Bandwidth
FRS-GMRS	462.5625	1	2 W/12.5 kHz	5 W/20 kHz
FRS-GMRS	462.5875	2	2 W/12.5 kHz	5 W/20 kHz
FRS-GMRS	462.6125	3	2 W/12.5 kHz	5 W/20 kHz
FRS-GMRS	462.6375	4	2 W/12.5 kHz	5 W/20 kHz
FRS-GMRS	462.6625	5	2 W/12.5 kHz	5 W/20 kHz
FRS-GMRS	462.6875	6	2 W/12.5 kHz	5 W/20 kHz
FRS-GMRS	462.7125	7	2 W/12.5 kHz	5 W/20 kHz
FRS-GMRS	467.5525	8	0.5 W/12.5 kHz	0.5W / 12.5kHz
FRS-GMRS	467.5875	9	0.5 W/12.5 kHz	0.5W / 12.5kHz
FRS-GMRS	467.6125	10	0.5 W/12.5 kHz	0.5W / 12.5kHz
FRS-GMRS	467.6375	11	0.5 W/12.5 kHz	0.5W / 12.5kHz
FRS-GMRS	467.6625	12	0.5 W/12.5 kHz	0.5W / 12.5kHz
FRS-GMRS	467.6875	13	0.5 W/12.5 kHz	0.5W / 12.5kHz
FRS-GMRS	467.7125	14	0.5 W/12.5 kHz	0.5W / 12.5kHz
GMRS	462.5500	15		Simplex or repeater output – 50 W maximum/20 kHz
GMRS	462.5750	16		Simplex or repeater output – 50 W maximum/20 kHz
GMRS	462.6000	17		Simplex or repeater output – 50 W maximum/20 kHz
GMRS	462.6250	18		Simplex or repeater output – 50 W maximum/20 kHz
GMRS	462.6500	19		Simplex or repeater output – 50 W maximum/20 kHz

Service	Frequency MHz	Channel #	FRS Max Power/Bandwidth	GMRS Max Power/Bandwidth
GMRS	462.6750	20		Simplex or repeater output – 50 W maximum/20 kHz
GMRS	462.7000	21		Simplex or repeater output – 50 W maximum/20 kHz
GMRS	462.7250	22		Simplex or repeater output – 50 W maximum/20 kHz

MURS (Multi-Use Radio Service) Frequencies

Note: For most ham radios in most circumstances, these are not – by the letter of the law – legal frequencies for us to use. In emergencies, however, we use whatever we need to use to prevent harm to life, limb or property.

Channel	Frequency	Maximum authorized bandwidth	Channel name
1	151.820 MHz	11.25 kHz	MURS 1
2	151.880 MHz	11.25 kHz	MURS 2
3	151.940 MHz	11.25 kHz	MURS 3
4	154.570 MHz	20.00 kHz	Blue Dot
5	154.600 MHz	20.00 kHz	Green Dot

NOAA Weather Radio Frequencies

162.400 MHz	162.425 MHz	162.450 MHz	162.475 MHz	162.500 MHz	162.525 MHz	162.550 MHz

Citizens Band Frequencies

Note: For most ham radios in most circumstances, these are not – by the letter of the law – legal frequencies for us to use. In emergencies, however, we use whatever we need to use to prevent harm to life, limb or property. Citizen's Band uses AM modulation and, rarely, SSB.

Channel	Frequency	Channel	Frequency	Channel	Frequency	Channel	Frequency
1	26.965 MHz	11	27.085 MHz	21	27.215 MHz	31	27.315 MHz
2	26.975 MHz	12	27.105 MHz	22	27.225 MHz	32	27.325 MHz
3	26.985 MHz	13	27.115 MHz	23	27.255 MHz	33	27.335 MHz
4	27.005 MHz	14	27.125 MHz	24	27.235 MHz	34	27.345 MHz
5	27.015 MHz	15	27.135 MHz	25	27.245 MHz	35	27.355 MHz
6	27.025 MHz	16	27.155 MHz	26	27.265 MHz	36	27.365 MHz
7	27.035 MHz	17	27.165 MHz	27	27.275 MHz	37	27.375 MHz
8	27.055 MHz	18	27.175 MHz	28	27.285 MHz	38	27.385 MHz
9	27.065 MHz	19	27.185 MHz	29	27.295 MHz	39	27.395 MHz
10	27.075 MHz	20	27.205 MHz	30	27.305 MHz	40	27.405 MHz

Aircraft VHF Frequencies

Frequency	Allocation
108.000 – 112.000 MHz	Aviation Terminal VOR and ILS Navigation
112.000 – 117.950 MHz	Aviation VOR Navigation
118.000 – 136.975 MHz	Aviation Communication
121.500 MHz	Aviation Distress (VHF Guard)
118.000 – 121.950 MHz	Air Traffic Control (Towers and ARTCC's)
121.3 MHz	Ground
121.7 MHz	Ground
121.9 MHz	Ground
121.975 – 122.675 MHz	FSS
122.0 MHz	En Route Flight Advisory Service (Flight Watch)
122.2 MHz	Universal FSS
122.700 MHz	Unicom (CTAF)
122.75 MHz	Air-to-Air (GA fixed wing)
122.800 MHz	Unicom (CTAF)
122.900 MHz	Unicom (CTAF)
122.950 MHz	Unicom (CTAF) for controlled airports
123.000 MHz	Unicom (CTAF)
123.025 MHz	Air-to-Air (GA helicopters)
123.050 MHz	Unicom (CTAF)
123.3 MHz	Air-to-Air (Gliders and hot air balloons)
123.5 MHz	Air-to-Air (Gliders and hot air balloons)

All VHF aircraft modulation is AM.

Marine VHF Frequencies

New Channel Number	Old Channel Number	Ship Transmit MHz	Ship Receive MHz	Use
1001	01A	156.050	156.050	Port Operations and Commercial, VTS. Available only in New Orleans / Lower Mississippi area.
1005	05A	156.250	156.250	Port Operations or VTS in the Houston, New Orleans and Seattle areas.
06	06	156.300	156.300	Intership Safety
1007	07A	156.350	156.350	Commercial. VDSMS
08	08	156.400	156.400	Commercial (Intership only). VDSMS
09	09	156.450	156.450	Boater Calling. Commercial and Non-Commercial. VDSMS
10	10	156.500	156.500	Commercial. VDSMS
11	11	156.550	156.550	Commercial. VTS in selected areas. VDSMS
12	12	156.600	156.600	Port Operations. VTS in selected areas.
13	13	156.650	156.650	Intership Navigation Safety (Bridge-to-bridge). Ships >20m length maintain a listening watch on this channel in US waters.
14	14	156.700	156.700	Port Operations. VTS in selected areas.
15	15	--	156.750	Environmental (Receive only). Used by Class C EPIRBs.
16	16	156.800	156.800	International Distress, Safety and Calling. Ships required to carry radio, USCG, and most coast stations maintain a listening watch on this channel. See our Watchkeeping Regulations page.
17	17	156.850	156.850	State & local govt maritime control
1018	18A	156.900	156.900	Commercial. VDSMS
1019	19A	156.950	156.950	Commercial. VDSMS
20	20	157.000	161.600	Port Operations (duplex)
1020	20A	157.000	157.000	Port Operations
1021	21A	157.050	157.050	U.S. Coast Guard only
1022	22A	157.100	157.100	Coast Guard Liaison and Maritime Safety Information Broadcasts. Broadcasts announced on channel 16.
1023	23A	157.150	157.150	U.S. Coast Guard only
24	24	157.200	161.800	Public Correspondence (Marine Operator). VDSMS
25	25	157.250	161.850	Public Correspondence (Marine Operator). VDSMS
26	26	157.300	161.900	Public Correspondence (Marine Operator). VDSMS
27	27	157.350	161.950	Public Correspondence (Marine Operator). VDSMS
28	28	157.400	162.000	Public Correspondence (Marine Operator). VDSMS

1063	63A	156.175	156.175	Port Operations and Commercial, VTS. Available only in New Orleans / Lower Mississippi area.
1065	65A	156.275	156.275	Port Operations
1066	66A	156.325	156.325	Port Operations
67	67	156.375	156.375	Commercial. Used for Bridge-to-bridge communications in lower Mississippi River. Intership only.
68	68	156.425	156.425	Non-Commercial. VDSMS
69	69	156.475	156.475	Non-Commercial. VDSMS
70	70	156.525	156.525	Digital Selective Calling (voice communications not allowed)
71	71	156.575	156.575	Non-Commercial. VDSMS
72	72	156.625	156.625	Non-Commercial (Intership only). VDSMS
73	73	156.675	156.675	Port Operations
74	74	156.725	156.725	Port Operations
77	77	156.875	156.875	Port Operations (Intership only)
1078	78A	156.925	156.925	Non-Commercial. VDSMS
1079	79A	156.975	156.975	Commercial. Non-Commercial in Great Lakes only. VDSMS
1080	80A	157.025	157.025	Commercial. Non-Commercial in Great Lakes only. VDSMS
1081	81A	157.075	157.075	U.S. Government only - Environmental protection operations.
1082	82A	157.125	157.125	U.S. Government only
1083	83A	157.175	157.175	U.S. Coast Guard only
84	84	157.225	161.825	Public Correspondence (Marine Operator). VDSMS
85	85	157.275	161.875	Public Correspondence (Marine Operator). VDSMS
86	86	157.325	161.925	Public Correspondence (Marine Operator). VDSMS
87	87	157.375	157.375	Public Correspondence (Marine Operator). VDSMS
88	88	157.425	157.425	Commercial, Intership only. VDSMS
AIS 1	AIS 1	161.975	161.975	Automatic Identification System (AIS)
AIS 2	AIS 2	162.025	162.025	Automatic Identification System (AIS)

Note: VDSMS (VHF Digital Small Message Services). Transmissions of short digital messages in accordance with RTCM Standard 12301.1 is allowed.

Frequencies are in MHz. Modulation is 16KF3E or 16KG3E.

Note that the four digit channel number beginning with the digits "10" indicates simplex use of the ship station transmit side of what had been an international duplex channel. These new channel numbers, now recognized internationally, were previously designated in the US by the two digit channel number ending with the letter "A". That is, the international channel 1005 has been designated in the US by channel 05A, and the US Coast Guard channel 1022 has been designated in the US as channel 22A. Four digit channels beginning with "20", sometimes shown by the two-digit channel number ending with the letter "B", indicates simplex use of the coast station transmit side of what normally was an international duplex channel. The U.S. does not currently use "B" or "20NN" channels in the VHF maritime band. Some VHF transceivers are equipped with an "International - U.S." switch to avoid conflicting use of these channels. See ITU Radio Regulation Appendix 18 and ITU-R M.1084-5 Annex 4, summarized here.

These new channel numbers should eventually begin to be displayed on new models of VHF marine radios.

Boaters should normally use channels listed as Non-Commercial. Channel 16 is used for calling other stations or for distress alerting. Channel 13 should be used to contact a ship when there is danger of collision. All ships of length 20m or greater are required to guard VHF channel 13, in addition to VHF channel 16, when operating within U.S. territorial waters. Users may be fined by the FCC for improper use of these channels.

NCDXF Beacon Stations

Slot	DX Entity	Call	CW	Location	Grid Square	Operator
1	United States	4U1UN		New York City	FN3Øas	UNRC
2	Canada	VE8AT		Eureka, Nunavut	EQ79ax	RAC/NARC
3	United States	W6WX		Mt. Umunhum, CA	CM97bd	NCDXF
4	Hawaii	KH6RS		Maui	BL10ts	Maui ARC
5	New Zealand	ZL6B		Masterton	RE78tw	NZART
6	Australia	VK6RBP		Rolystone	OF87av	WIA
7	Japan	JA2IGY		Mt. Asama	PM84jk	JARL
8	Russia	RR9O		Novosibirsk	NO14kx	SRR
9	Hong Kong	VR2B		Hong Kong	OL72bg	HARTS
10	Sri Lanka	4S7B		Colombo	MJ96wv	RSSL
11	South Africa	ZS6DN		Pretoria	KG44dc	ZS6DN
12	Kenya	5Z4B		Kariobangi	KI88ks	ARSK
13	Israel	4X6TU		Tel Aviv	KM72jb	IARC
14	Finland	OH2B		Lohja	KP2Ø	SRAL
15	Madeira	CS3B		São Jorge	IM12mt	Delegação
16	Argentina	LU4AA		Buenos Aires	GFØ5tj	RCA
17	Peru	OA4B		Lima	FH17mw	RCP
18	Venezuela	YV5B		Caracas	FJ69cc	RCV

The slot indicates the order of transmission. Each beacon transmission is staggered by 10 seconds so that no two beacons are on the same frequency at the same time.

The beacons transmit on 5 frequencies: 14.100, 18.110, 21.150, 24.930, 28.200 MHz in a 3-minute cycle so that no two beacons transmit at the same time on the same frequency. The timing is shown on the Beacon Transmission Schedule page, https://www.ncdxf.org/beacon/index.

NATO Phonetic Alphabet

Letter	Phonetic	Pronunciation	Letter	Phonetic	Pronunciation
A	Alfa	AL-fah	S	Sierra	see-AIR-ah
B	Bravo	BRAH-voh	T	Tango	TANG-go
C	Charlie	CHAR-lee	U	Uniform	YOU-nee-form
D	Delta	DEL-tah	V	Victor	VICK-tor
E	Echo	ECK-oh	W	Whiskey	WISS-key
F	Foxtrot	FOKS-trot	X	Xray	ECKS-ray
G	Golf	GOLF	Y	Yankee	YANG-key
H	Hotel	hoh-TELL	Z	Zulu	ZOO-loo
I	India	IN-dee-ah	Ø	Zero	Zero
J	Juliett	JEW-lee-ett	1	One	One
K	Kilo	KEY-loh	2	Two	Two
L	Lima	LEE-mah	3	Three	Three
M	Mike	MIKE	4	Four	Four
N	November	no-VEM-ber	5	Five	Five
O	Oscar	OSS-car	6	Six	Six
P	Papa	pah-PAH	7	Seven	Seven
Q	Quebec	keh-BECK	8	Eight	Eight
R	Romeo	ROW-me-oh	9	Niner	NINE-er

International Morse Code

A	·—	Q	——·—	6	—····
B	—···	R	·—·	7	——···
C	—·—·	S	···	8	———··
D	—··	T	—	9	————·
E	·	U	··—	,	——··——
F	··—·	V	···—	.	·—·—·—
G	——·	W	·——	?	··——··
H	····	X	—··—	:	———···
I	··	Y	—·——	;	—·—·—·
J	·———	Z	——··	/	—··—·
K	—·—	Ø	—————	-	—····—
L	·—··	1	·————	_	·—·—·—
M	——	2	··———	(	—·——·—
N	—·	3	···——	)	—·——·—
O	———	4	····—	+	·—·—·
P	·——·	5	·····	-	—····—

Call Sign Formats

	US Amateur Call Signs	
Type	Issued To	Examples
2 x 3	Technician, General (Extra has option to keep 2 x 3 call sign)	AB1BBB, KB2BBB, NB3BBB, WB4BBB
1 x 3	Technician, General (Extra has option to keep 1 x 3 call sign)	K5BBB, N6BBB, W7BBB
2 x 2	Amateur Extra, grandfathered Advanced	AB8BB, KB9BB, NB0BB, WB1BB
1 x 2	Amateur Extra, grandfathered Advanced	A2BB, N3BB, W4BB
1 x 1	Special Event Stations	K5B, N6B, W7B
	US call signs begin with AA – AL, K, N, or W.	

US Call Sign Regions

Callsign Numeral	Region	Callsign Numeral	Region
1	Connecticut (CT), Maine (ME), Massachusetts (MA), New Hampshire (NH), Rhode Island (RI), Vermont (VT)	6	California (CA), Hawaii (HI), KH6, Pacific Islands
2	New Jersey (NJ), New York (NY)	7	Alaska (AK), KL7, Arizona (AZ), Idaho (ID), Montana (MT), Nevada (NV), Oregon (OR), Utah (UT), Washington (WA) Wyoming (WY)
3	Delaware (DE), Maryland (MD), Pennsylvania (PA)	8	Michigan (MI) Ohio (OH), West Virginia (WV)
4	Alabama (AL), Florida (FL), Georgia (GA), Kentucky (KY), North Carolina (NC), Puerto Rico (PR), South Carolina (SC), Tennessee (TN), US Virgin Islands, Virginia (VA)	9	Illinois (IL), Indiana (IN), Wisconsin (WI)
5	Arkansas (AR), Louisiana (LA), Mississippi (MS), New Mexico (NM), Oklahoma (OK), Texas (TX)	Ø	Colorado (CO), Iowa (IA) Kansas (KS), Minnesota (MN) Missouri (MO), Nebraska (NE) North Dakota (ND), South Dakota (SD)

Allocations of World Call Sign Prefixes

Prefix	Country Name	Continent	ITU Zone	CQ Zone	Time Zone vs UTC	Latitude	Longitude
~~*	Blenheim Reef	Af	41	39	+5	7S	72E
~~*	Geyser Reef	Af	53	39	+3	12S	46E
1A	SMO Malta	Eu	28	15	+1	42N	13E
1M*	Minerva Reef	Oc	62	32	-12	24S	179W
1S	Spratly Is	As	50	26	+7	9N	112E
2D	Isle of Man	Eu	27	14	+0	54N	4W
2E	England	Eu	27	14	+0	52N	0W
2I	No Ireland	Eu	27	14	+0	55N	6W
2J	Jersey	Eu	27	14	+0	49N	2W
2M	Scotland	Eu	27	14	+0	57N	2W
2U	Guernsey	Eu	27	14	+0	49N	3W
2W	Wales	Eu	27	14	+0	52N	3W
3A	Monaco	Eu	27	14	+1	44N	8E
3B6	Agalega & St Brandon	Af	53	39	+4	10S	57E
3B7	Agalega & St Brandon	Af	53	39	+4	10S	57E
3B8	Mauritius	Af	53	39	+4	20S	58E
3B9	Rodriguez Is	Af	53	39	+4	20S	63E
3C	Equatorial Guinea	Af	47	36	-1	4N	9E
3C0	Annobon Is	Af	52	36	-1	1S	6E
3D2	Conway Reef	Oc	56	32	+12	22S	175E
3D2	Fiji	Oc	56	32	+12	18S	178E
3D2	Rotuma	Oc	56	32	+12	13S	177E
3D6	Swaziland	Af	57	38	+2	26S	31E
3DA	Swaziland	Af	57	38	+2	26S	31E
3H-3U	China	As	33,42-44	23,24	+8	40N	116E
3V	Tunisia	Af	37	33	+1	37N	10E
3W	Vietnam	As	49	26	+7	11N	107E
3X	Guinea	Af	46	35	+0	10N	14W
3Y	Bouvet	Af	67	38	+0	54S	3E
3Y	Peter I	An	72	12	-6	69S	91W
3Y	Antarctica	An	67,69-74	12,13,29,30,32,38,39	+0	90S	0W
4J	Azerbaijan	As	29	21	+4	40N	50E
4L	Georgia	As	29	21	+4	42N	45E

Prefix	Country Name	Continent	ITU Zone	CQ Zone	Time Zone vs UTC	Latitude	Longitude
4N5	Macedonia	Eu	28	15	+1	42N	22E
4O	Montenegro	Eu	28	15	+1	42N	19E
4S	Sri Lanka	As	41	22	+5.5	7N	80E
4U	ITU Geneva	Eu	28	14	+1	46N	6E
4U	UN HQ	NA	08	05	-5	41N	74W
4W	Timor Leste	Oc	54	28	+8	9S	126E
4W*	Yemen Arab Rep	As	39	21	+3	15N	44E
4X	Israel	As	39	20	+2	32N	35E
4X1	Palestine	As	39	20	+2	32N	35E
4Y	International Civil Aviation Organization						
5A	Libya	Af	38	34	+2	33N	13E
5B	Cyprus	As	39	20	+3	35N	33E
5H	Tanzania	Af	53	37	+3	7S	39E
5N	Nigeria	Af	46	35	+1	6N	3E
5R	Madagascar	Af	53	39	+3	19S	48E
5T	Mauritania	Af	46	35	-1	18N	16W
5U	Niger	Af	46	35	+1	14N	2W
5V	Togo	Af	46	35	+0	6N	1E
5W	Samoa	Oc	62	32	-11	14S	172W
5X	Uganda	Af	48	37	+3	0N	33E
5Z	Kenya	Af	48	37	+3	2S	37E
6W	Senegal	Af	46	35	+0	15N	18W
6Y	Jamaica	NA	11	08	-5	18N	77W
7JI	Okino Tori-shima	As	45	27	+10	30N	140E
7O	Yemen	As	39	21	+3	13N	45E
7O	Kamaran Is	As	39	21	+3	15N	43E
7O*	PDR Yemen	As	39	21	+3	13N	45E
7P	Lesotho	Af	57	38	+2	29S	27E
7Q	Malawi	Af	53	37	+2	14S	34E
7X	Algeria	Af	37	33	+0	37N	3E
8J1	Antarctica	An	67,69-74	12,13,29,30,32,38,39	+0	90S	0W
8P	Barbados	NA	11	08	-4	13N	60W
8Q	Maldives	As,Af	41	22	+5	4N	73E
8R	Guyana	SA	12	09	-3.75	6N	58W
8Z4*	S Arabia/Iraq NZ	As	39	21	+3	29N	46E

Prefix	Country Name	Continent	ITU Zone	CQ Zone	Time Zone vs UTC	Latitude	Longitude
8Z5*	Kuwait/S Arabia NZ	As	39	21	+3	29N	48E
9A	Croatia	Eu	28	15	+1	46N	16E
9G	Ghana	Af	46	35	+0	5N	0W
9H	Malta	Eu	28	15	+1	36N	15E
9J	Zambia	Af	53	36	+2	15S	28E
9K	Kuwait	As	39	21	+3	29N	48E
9K3	Kuwait/S Arabia NZ	As	39	21	+3	29N	48E
9L	Sierra Leone	Af	46	35	+0	9N	13W
9M0	Spratly Is	As	50	26	+7	9N	112E
9M2	West Malaysia	As	54	28	+7.5	3N	102E
9M2	Malaya	As	54	28	+7.5	3N	102E
9M4							
9M6	East Malaysia	Oc	54	28	+8	2N	110E
9N	Nepal	As	42	22	+5.75	28N	85E
9Q	Dem Rep Congo (Zaire)	Af	52	36	+1	4S	15E
9S4*	Saar	Eu	28	14	+1	49N	7E
9U	Burundi	Af	52	36	+3	3S	29E
9U5*	Ruanda-Urundi	Af	52	36	+3	3S	30E
9V	Singapore	As	54	28	+7.5	1N	104E
9X	Rwanda	Af	52	36	+3	2S	30E
9Y	Trinidad & Tobago	SA	11	09	-4	11N	62W
A1*	Abu Ail, Jabal at Tair	As	39	21	+2	14N	43E
A2	Botswana	Af	57	38	+2	25S	26E
A3	Tonga	Oc	62	32	+13	21S	175W
A4	Oman	As	39	21	+4	24N	59E
A5	Bhutan	As	41	22	+5.5	28N	90E
A6	United Arab Emirates	As	39	21	+4	24N	54E
A7	Qatar	As	39	21	+4	25N	52E
A9	Bahrain	As	39	21	+4	26N	51E
AC3*	Sikkim	As	41	22	+5.5	27N	89E
AC4*	Tibet	As	41	23	+6	30N	92E
AH0	Mariana Is	Oc	64	27	+10	15N	146E
AH1	Baker & Howland Is	Oc	61	31	-12	0N	176W

Prefix	Country Name	Continent	ITU Zone	CQ Zone	Time Zone vs UTC	Latitude	Longitude
AH2	Guam	Oc	64	27	+10	13N	145E
AH3	Johnston Is	Oc	61	31	-11	17N	170W
AH4	Midway Is	Oc	61	31	-11	28N	177W
AH5	Palmyra, Jarvis Is	Oc	61,62	31	-11	6N	162W
AH5K	Kingman Reef	Oc	61	31	-11	6N	162W
AH6	Hawaii	Oc	61	31	-10	21N	158W
AH7	Hawaii	Oc	61	31	-10	21N	158W
AH7K	Kure Is	Oc	61	31	-11	29N	178W
AH8	Am Samoa	Oc	62	32	-11	14S	171W
AH8S	Swain's Island	Oc	62	32	-11	11S	171W
AH9	Wake Is	Oc	65	31	+12	19N	167E
AL	Alaska	NA	01,02	01	-8	58N	134W
AP	Pakistan	As	41	21	+5	34N	73E
AT0	Antarctica	An	67,69-74	12,13,29,30,32,38,39	+0	90S	0W
BA-BZ	China	As	33,42-44	23,24	+8	40N	116E
BM-BQ	Taiwan	As	44	24	+8	25N	122E
BQ9	Pratas Is	As	44	24	+8	21N	116E
BS7	Scarborough Reef	As	50	27	+8	15N	118E
BU	Taiwan	As	44	24	+8	25N	122E
BV	Taiwan	As	44	24	+8	25N	122E
BV9	Pratas Is	As	44	24	+8	21N	116E
BW	Taiwan	As	44	24	+8	25N	122E
BX	Taiwan	As	44	24	+8	25N	122E
BY	China	As	33,42-44	23,24	+8	40N	116E
C2	Nauru	Oc	65	31	+11.5	1S	167E
C3	Andorra	Eu	27	14	+1	43N	2E
C5	Gambia	Af	46	35	+0	13N	17W
C6	Bahamas	NA	11	08	-5	25N	77W
C7	World Meteorological Organization						
C9	Mozambique	Af	53	37	+2	26S	33E
C9*	Manchuria	As	33	24	+8.5	46N	127E
CE	Chile	SA	14,16	12	-4	33S	71W
CE0X	San Felix	SA	14	12	-5	26S	80W
CE0Y	Easter Is	SA	63	12	-7	27S	109W
CE0Z	Juan Fernandez	SA	14	12	-4	34S	79W

Prefix	Country Name	Continent	ITU Zone	CQ Zone	Time Zone vs UTC	Latitude	Longitude
CE9	Antarctica	An	67,69-74	12,13,29,30,32,38,39	+0	90S	0W
CE9	So Shetland Is	SA	73	13	-4	62S	58W
CN	Morocco	Af	37	33	+0	34N	7W
CN2*	Tangier	Af	37	33	+0	36N	8W
CO	Cuba	NA	11	08	-5	23N	82W
CP	Bolivia	SA	12,14	10	-4	17S	68W
CR8*	Damao, Diu	As	41	22	+5.5	21N	71E
CR8*	Goa	As	41	22	+5.5	16N	74E
CR8*	Portuguese Timor	Oc	54	28	+8	9S	126E
CT	Portugal	Eu	37	14	+0	39N	9W
CT3	Madeira Is	Af	36	33	-1	33N	17W
CU	Azores	Eu	36	14	-1	38N	26W
CX	Uruguay	SA	14	13	-3	35S	56W
CY0	Sable Is	NA	09	05	-5	44N	60W
CY9	St Paul Is	NA	09	05	-5	47N	60W
D2	Angola	Af	52	36	+1	9S	13E
D4	Cape Verde	Af	46	35	-2	15N	23W
D6	Comoros	Af	53	39	+3	12S	43E
DL	Germany	Eu	28	14	+1	53N	13E
DL*	Germany	Eu	28	14	+1	52N	7E
DM	East Germany	Eu	28	14	+1	53N	13E
DP0	Antarctica	An	67,69-74	12,13,29,30,32,38,39	+0	90S	0W
DU	Philippines	Oc	50	27	+8	15N	121E
E3	Eritrea	Af	48	37	+3	15N	39E
E4	Palestine	As	39	20	+2	32N	34E
E5	No Cook Is	Oc	62,63	32	-10.5	10S	161W
E5	So Cook Is	Oc	63	32	-10.5	22S	158W
E6	Niue	Oc	62	32	-11	19S	170W
E7	Bosnia-Hercegovina	Eu	28	15	+1	44N	18E
EA	Spain	Eu	37	14	+1	40N	4W
EA6	Balearic Is	Eu	37	14	+1	38N	3E
EA8	Canary Is	Af	36	33	+0	28N	15W
EA9	Ceuta & Melilla	Af	37	33	+1	36N	5W
EA9*	Ifni	Af	37	33	+0	29N	10W
EI	Ireland	Eu	27	14	+0	53N	6W
EK	Armenia	As	29	21	+4	40N	45E

Prefix	Country Name	Continent	ITU Zone	CQ Zone	Time Zone vs UTC	Latitude	Longitude
EL	Liberia	Af	46	35	-0.75	6N	11W
EP	Iran	As	40	21	+3.5	36N	51E
ER	Moldova	Eu	29	16	+3	47N	29E
ES	Estonia	Eu	29	15	+2	59N	25E
ET	Ethiopia	Af	48	37	+3	9N	39E
ET2	Eritrea	Af	48	37	+3	15N	39E
EV	Belarus	Eu	29	16	+2	54N	28E
EX	Kyrgyzstan	As	30,31	17	+6	43N	75E
EY	Tajikistan	As	30	17	+6	39N	69E
EZ	Turkmenistan	As	30	17	+5	38N	58E
F	France	Eu	27	14	+1	49N	2E
FB8	Comoros	Af	53	39	+3	12S	43E
FF*	French W Africa	Af	46	35	+0	15N	18W
FG	Guadeloupe	NA	11	08	-4	16N	62W
FH	Mayotte	Af	53	39	+3	13S	45E
FH*	Comoros	Af	53	39	+3	12S	43E
FI8*	Fr Indo China	As	49	26	+7	11N	107E
FJ	St Barthelemy	NA	11	08	-4	18N	63W
FK	Chesterfield Is	Oc	55	30	+11	20S	158E
FK	New Caledonia	Oc	56	32	+11	22S	167E
FM	Martinique	NA	11	08	-4	15N	61W
FN8*	French India	As	41	22	+5.5	12N	80E
FO	French Polynesia	Oc	63	32	-10	18S	150W
FO0	Austral Is	Oc	63	32	-10	23S	149W
FO0	Clipperton Is	NA	10	07	-7	10N	109W
FO0	Marquesas Is	Oc	63	31	-10	9S	140W
FP	St Pierre & Miquelon	NA	09	05	-4	47N	56W
FQ8*	Fr Equatorial Africa	Af	47,52	36	+1	5N	18E
FR	Reunion	Af	53	39	+4	21S	55E
FR/G	Glorioso Is	Af	53	39	+3	12S	47E
FR/J	Juan de Nova, Europa	Af	53	39	+3	17S	43E
FR/T	Tromelin	Af	53	39	+4	16S	54E
FS	St Martin	NA	11	08	-4	18N	63W
FT_E	Juan de Nova,	Af	53	39	+3	17S	43E

Prefix	Country Name	Continent	ITU Zone	CQ Zone	Time Zone vs UTC	Latitude	Longitude
	Europa						
FT_G	Glorioso Is	Af	53	39	+3	12S	47E
FT_J	Juan de Nova, Europa	Af	53	39	+3	17S	43E
FT_T	Tromelin	Af	53	39	+4	16S	54E
FT_W	Crozet	Af	68	39	+3	46S	52E
FT_X	Kerguelen Is	Af	68	39	+5	50S	70E
FT_Y	Antarctica	An	67,69-74	12,13,29,30,32,38,39	+0	90S	0W
FT_Z	Amsterdam & St Paul Is	Af	68	39	+5	38S	78E
FW	Wallis & Futuna Is	Oc	62	32	-10.5	14S	172W
FY	French Guiana	SA	12	09	-4	5N	52W
G	England	Eu	27	14	+0	52N	0W
GC	Wales	Eu	27	14	+0	52N	3W
GD	Isle of Man	Eu	27	14	+0	54N	4W
GH	Jersey	Eu	27	14	+0	49N	2W
GI	No Ireland	Eu	27	14	+0	55N	6W
GJ	Jersey	Eu	27	14	+0	49N	2W
GM	Scotland	Eu	27	14	+0	57N	2W
GN	No Ireland	Eu	27	14	+0	55N	6W
GP	Guernsey	Eu	27	14	+0	49N	3W
GS	Scotland	Eu	27	14	+0	57N	2W
GT	Isle of Man	Eu	27	14	+0	54N	4W
GU	Guernsey	Eu	27	14	+0	49N	3W
GW	Wales	Eu	27	14	+0	52N	3W
GX	England	Eu	27	14	+0	52N	0W
GZ	Scotland	Eu	27	14	+0	57N	2W
H4	Solomon Is	Oc	51	28	+11	9S	160E
H40	Temotu	Oc	51	28	+11	11S	166E
HA	Hungary	Eu	28	15	+1	48N	19E
HB	Switzerland	Eu	28	14	+1	47N	7E
HB0	Liechtenstein	Eu	28	14	+1	47N	10E
HC	Ecuador	SA	12	10	-5	0N	79W
HC8	Galapagos Is	SA	12	10	-6	1S	90W
HF0	So Shetland Is	SA	73	13	-4	62S	58W
HH	Haiti	NA	11	08	-5	19N	72W
HI	Dominican Rep	NA	11	08	-5	18N	70W

Prefix	Country Name	Continent	ITU Zone	CQ Zone	Time Zone vs UTC	Latitude	Longitude
HK	Colombia	SA	12	09	-5	5N	74W
HK0	Malpelo Is	SA	12	09	-5	4N	82W
HK0	San Andres & Providencia	NA	11	07	-6	13N	82W
HK0*	Bajo Nuevo	NA	11	08	-5	16N	79W
HK0*	Serrana Bnk, Roncador Cay	NA	11	07	-5	14N	80W
HL	So Korea	As	44	25	+9	38N	127E
HP	Panama	NA	11	07	-5	9N	80W
HR	Honduras	NA	11	07	-6	14N	87W
HS	Thailand	As	49	26	+7	14N	101E
HV	Vatican	Eu	28	15	+1	42N	13E
HZ	Saudi Arabia	As	39	21	+3	25N	47E
I	Italy	Eu,Af	28,37	15,33	+1	42N	12E
I1*	Trieste	Eu	28	15	+1	46N	14E
I5*	Ital Somaliland	Af	48	37	+3	2N	46E
IS	Sardinia	Eu	28	15	+1	39N	9E
J2	Djibouti	Af	48	37	+3	12N	43E
J2/A	Abu Ail, Jabal at Tair	As	39	21	+2	14N	43E
J3	Grenada	NA	11	08	-4	12N	62W
J5	Guinea-Bissau	Af	46	35	-1	12N	16W
J6	St Lucia	NA	11	08	-4	14N	61W
J7	Dominica	NA	11	08	-4	15N	61W
J8	St Vincent	NA	11	08	-4	13N	61W
JA	Japan	As	45	25	+9	36N	140E
JD	Minami Torishima	Oc	90	27	+10	24N	154E
JD	Ogasawara	As	45	27	+10	28N	142E
JD1*	Okino Tori-shima	As	45	27	+10	30N	140E
JR6	Okinawa (Ryukyu)	As	45	25	+8	26N	128E
JT	Mongolia	As	32,33	23	+7.5	48N	107E
JW	Svalbard	Eu	18	40	+1	78N	16E
JX	Jan Mayen	Eu	18	40	-1	71N	9W
JY	Jordan	As	39	20	+2	32N	36E
JZ0*	Neth New Guinea	Oc	51	28	+10	10S	147E
K	United States	NA	06-08	03-05	-5	39N	77W

Prefix	Country Name	Continent	ITU Zone	CQ Zone	Time Zone vs UTC	Latitude	Longitude
KA6	Okinawa (Ryukyu)	As	45	25	+8	26N	128E
KC4	Antarctica	An	67,69-74	12,13,29,30,32,38,39	+0	90S	0W
KC6	Palau	Oc	64	27	+10	7N	134E
KC6	Micronesia	Oc	65	27	+11	7N	158E
KG4	Guantanamo Bay	NA	11	08	-5	20N	75W
KH0	Mariana Is	Oc	64	27	+10	15N	146E
KH1	Baker & Howland Is	Oc	61	31	-12	0N	176W
KH2	Guam	Oc	64	27	+10	13N	145E
KH3	Johnston Is	Oc	61	31	-11	17N	170W
KH4	Midway Is	Oc	61	31	-11	28N	177W
KH5	Palmyra, Jarvis Is	Oc	61,62	31	-11	6N	162W
KH5K*	Kingman Reef	Oc	61	31	-11	6N	162W
KH6	Hawaii	Oc	61	31	-10	21N	158W
KH7K	Kure Is	Oc	61	31	-11	29N	178W
KH7K	Hawaii	Oc	61	31	-10	21N	158W
KH8	Am Samoa	Oc	62	32	-11	14S	171W
KH8S	Swain's Island	Oc	62	32	-11	11S	171W
KH9	Wake Is	Oc	65	31	+12	19N	167E
KL	Alaska	NA	01,02	01	-8	58N	134W
KP1	Navassa Is	NA	11	08	-5	18N	75W
KP2	Virgin Is	NA	11	08	-4	18N	65W
KP3	Serrana Bnk, Roncador Cay	NA	11	07	-5	14N	80W
KP3	Puerto Rico	NA	11	08	-4	18N	66W
KP4	Puerto Rico	NA	11	08	-4	18N	66W
KP5	Desecheo Is	NA	11	08	-4	18N	68W
KR6*	Okinawa (Ryukyu)	As	45	25	+8	26N	128E
KR8	Okinawa (Ryukyu)	As	45	25	+8	26N	128E
KS4*	Swan Is	NA	11	07	-6	17N	84W
KS4*	Serrana Bnk, Roncador Cay	NA	11	07	-5	14N	80W
KX6	Marshall Is	Oc	65	31	+12	7N	171E
KZ5*	Canal Zone	NA	11	07	-5	9N	80W
LA	Norway	Eu	18	14	+1	60N	11E

Prefix	Country Name	Continent	ITU Zone	CQ Zone	Time Zone vs UTC	Latitude	Longitude
LU	Argentina	SA	14,16	13	-3	35S	58W
LU_Z	Antarctica	An	67,69-74	12,13,29,30,32,38,39	+0	90S	0W
LU_Z	So Georgia Is	SA	73	13	-1.5	54S	37W
LU_Z	So Orkney Is	SA	73	13	-3	61S	45W
LU_Z	So Sandwich Is	SA	73	13	-3	59S	27W
LU_Z	So Shetland Is	SA	73	13	-4	62S	58W
LX	Luxembourg	Eu	27	14	+1	50N	6E
LY	Lithuania	Eu	29	15	+2	55N	25E
LZ	Bulgaria	Eu	28	20	+2	43N	23E
MC	Wales	Eu	27	14	+0	52N	3W
MD	Isle of Man	Eu	27	14	+0	54N	4W
MH	Jersey	Eu	27	14	+0	49N	2W
MI	No Ireland	Eu	27	14	+0	55N	6W
MJ	Jersey	Eu	27	14	+0	49N	2W
MM	Scotland	Eu	27	14	+0	57N	2W
MN	No Ireland	Eu	27	14	+0	55N	6W
MP	Guernsey	Eu	27	14	+0	49N	3W
MS	Scotland	Eu	27	14	+0	57N	2W
MT	Isle of Man	Eu	27	14	+0	54N	4W
MU	Guernsey	Eu	27	14	+0	49N	3W
MW	Wales	Eu	27	14	+0	52N	3W
mx	England	Eu	27	14	+0	52N	0W
MX	England	Eu	27	14	+0	52N	0W
MZ	Scotland	Eu	27	14	+0	57N	2W
NH0	Mariana Is	Oc	64	27	+10	15N	146E
NH1	Baker & Howland Is	Oc	61	31	-12	0N	176W
NH2	Guam	Oc	64	27	+10	13N	145E
NH3	Johnston Is	Oc	61	31	-11	17N	170W
NH4	Midway Is	Oc	61	31	-11	28N	177W
NH5	Palmyra, Jarvis Is	Oc	61,62	31	-11	6N	162W
NH5K	Kingman Reef	Oc	61	31	-11	6N	162W
NH6	Hawaii	Oc	61	31	-10	21N	158W
NH7	Hawaii	Oc	61	31	-10	21N	158W
NH7K	Kure Is	Oc	61	31	-11	29N	178W
NH8	Am Samoa	Oc	62	32	-11	14S	171W
NH8S	Swain's Island	Oc	62	32	-11	11S	171W

Prefix	Country Name	Continent	ITU Zone	CQ Zone	Time Zone vs UTC	Latitude	Longitude
NH9	Wake Is	Oc	65	31	+12	19N	167E
NL	Alaska	NA	01,02	01	-8	58N	134W
NP1	Navassa Is	NA	11	08	-5	18N	75W
NP2	Virgin Is	NA	11	08	-4	18N	65W
NP3	Puerto Rico	NA	11	08	-4	18N	66W
NP4	Puerto Rico	NA	11	08	-4	18N	66W
NP5	Desecheo Is	NA	11	08	-4	18N	68W
OA	Peru	SA	12	10	-5	12S	78W
OD	Lebanon	As	39	20	+2	34N	36E
OE	Austria	Eu	28	15	+1	48N	16E
OH	Finland	Eu	18	15	+2	60N	25E
OH0	Aland Is	Eu	18	15	+2	60N	20E
OJ0	Market Reef	Eu	18	15	+2	60N	19E
OK	Czech Republic	Eu	28	15	+1	50N	15E
OK*	Czechoslovakia	Eu	28	15	+1	50N	15E
OM	Slovakia	Eu	28	15	+1	48N	17E
ON	Belgium	Eu	27	14	+1	51N	4E
OR4	Antarctica	An	67,69-74	12,13,29,30,32,38,39	+0	90S	0W
OW	Faroe Is	Eu	18	14	+0	62N	7W
OX	Greenland	NA	05,75	40	-3	64N	52W
OY	Faroe Is	Eu	18	14	+0	62N	7W
OZ	Denmark	Eu	18	14	+1	56N	13E
P2	Papua New Guinea	Oc	51	28	+10	10S	147E
P2*	Papua Terr	Oc	51	28	+10	10S	147E
P2*	Terr New Guinea	Oc	51	28	+10	10S	147E
P4	Aruba	SA	11	09	-4	13N	70W
P5	No Korea	As	44	25	+9	39N	126E
PA	Netherlands	Eu	27	14	+1	52N	5E
PJ2	Curacao	SA	11	09	-4	12N	69W
PJ2*	Neth Antilles	SA	11	09	-4	12N	69W
PJ4	Bonaire	SA	11	09	-4	12N	68W
PJ4	Neth Antilles	SA	11	09	-4	12N	69W
PJ5	Saba & St Eustatius	NA	11	08	-4	18N	63W
PJ5*	St Maarten, Saba, St Eus	NA	11	08	-4	18N	63W
PJ6	Saba & St	NA	11	08	-4	18N	63W

Prefix	Country Name	Continent	ITU Zone	CQ Zone	Time Zone vs UTC	Latitude	Longitude
	Eustatius						
PJ6-8	St Maarten, Saba, St Eus	NA	11	08	-4	18N	63W
PJ7	St Maarten	NA	11	08	-4	18N	63W
PJ9	Neth Antilles	SA	11	09	-4	12N	69W
PK1*	Java	Oc	54	28	+7.5	6S	107E
PK2-3	Java	Oc	54	28	+7.5	6S	107E
PK4*	Sumatra	Oc	54	28	+7	1S	100E
PK5*	Netherlands Borneo	Oc	54	28	+8	3S	115E
PK6*	Celebe & Molucca Is	Oc	54	28	+8	5S	119E
PY	Brazil	SA	12,13,15	11	-3	16S	48W
PY0F	Fernando de Noronha	SA	13	11	-2	4S	32W
PY0P	St Peter & St Paul Rocks	SA	13	11	-2	1N	29W
PY0T	Trindade & Martin Vaz Is	SA	15	11	-2	21S	29W
PY0ZF	Fernando de Noronha	SA	13	11	-2	4S	32W
PY0ZP	St Peter & St Paul Rocks	SA	13	11	-2	1N	29W
PY0ZT	Trindade & Martin Vaz Is	SA	15	11	-2	21S	29W
PZ	Surinam	SA	12	09	-3.5	6N	55W
R	Russia	Eu	19,20,29,30	16	+3	56N	37E
R8	Russia (Asiatic)	As	20-26,30-35,75	16-19,23	+7	52N	104E
R9-0	Russia (Asiatic)	As	20-26,30-35,75	16-19,23	+7	52N	104E
RA2	Kaliningrad	Eu	29	15	+2	55N	21E
RI1AN	Antarctica	An	67,69-74	12,13,29,30,32,38,39	+0	90S	0W
RI1AN	So Shetland Is	SA	73	13	-4	62S	58W
RI1F	Franz Josef Land	Eu	75	40	+3	81N	48E
RI1M*	Malyj Vysotskij Is	Eu	29	16	+3	61N	29E
S0	Western Sahara	Af	46	33	+0	27N	13W
S2	Bangladesh	As	41	22	+6	24N	90E
S5	Slovenia	Eu	28	15	+1	46N	15E
S7	Seychelles	Af	53	39	+4	5S	55E

Prefix	Country Name	Continent	ITU Zone	CQ Zone	Time Zone vs UTC	Latitude	Longitude
S9	Sao Tome & Principe	Af	47	36	+0	0N	7E
SH1	Zanzibar	Af	53	37	+3	7S	39E
SM	Sweden	Eu	18	14	+1	59N	18E
SP	Poland	Eu	28	15	+1	52N	21E
ST	Sudan	Af	48	34	+2	16N	33E
ST0*	Southern Sudan	Af	48	34	+2	5N	32E
SU	Egypt	Af,As	38	34	+2	31N	31E
SV	Greece	Eu	28	20	+2	38N	24E
SV/A	Mt Athos	Eu	28	20	+2	40N	24E
SV5	Dodecanese	Eu	28	20	+2	36N	28E
SV9	Crete	Eu	28	20	+2	36N	24E
SY2	Mt Athos	Eu	28	20	+2	40N	24E
T2	Tuvalu	Oc	65	31	+12	9S	179E
T30	West Kiribati	Oc	65	31	+12	1S	173E
T31	Central Kiribati	Oc	62	31	+12	4S	171W
T32	East Kiribati	Oc	61,63	31	+12	2N	158W
T33	Banaba	Oc	65	31	+11.5	1S	170E
T5	Somalia	Af	48	37	+3	2N	46E
T6	Afghanistan	As	40	21	+4.5	35N	69E
T7	San Marino	Eu	28	15	+1	44N	12E
T8	Palau	Oc	64	27	+10	7N	134E
T9	Bosnia-Hercegovina	Eu	28	15	+1	44N	18E
TA	Turkey	As,Eu	39	20	+2	40N	33E
TF	Iceland	Eu	17	40	+0	64N	22W
TG	Guatemala	NA	11	07	-6	16N	92W
TI	Costa Rica	NA	11	07	-6	10N	84W
TI9	Cocos Is	NA	11	07	-6	6N	87W
TJ	Cameroon	Af	47	36	+1	4N	12E
TK	Corsica	Eu	28	15	+1	42N	9E
TL	Central African Rep	Af	47	36	+1	5N	19E
TN	Congo	Af	52	36	+1	4S	15E
TO	Guadeloupe	NA	11	08	-4	16N	62W
TO	St Barthelemy	NA	11	08	-4	18N	63W
TO	Martinique	NA	11	08	-4	15N	61W
TO	Reunion	Af	53	39	+4	21S	55E

Prefix	Country Name	Continent	ITU Zone	CQ Zone	Time Zone vs UTC	Latitude	Longitude
TO	St Martin	NA	11	08	-4	18N	63W
TO	Glorioso Is	Af	53	39	+3	12S	47E
TO	Juan de Nova, Europa	Af	53	39	+3	17S	43E
TO	Tromelin	Af	53	39	+4	16S	54E
TO	French Guiana	SA	12	09	-4	5N	52W
TR	Gabon	Af	52	36	+1	1N	10E
TT	Chad	Af	47	36	+1	12N	15E
TU	Ivory Coast	Af	46	35	+0	7N	5W
TX	Mayotte	Af	53	39	+3	13S	45E
TX	Chesterfield Is	Oc	55	30	+11	20S	158E
TX	New Caledonia	Oc	56	32	+11	22S	167E
TX	French Polynesia	Oc	63	32	-10	18S	150W
TX	Austral Is	Oc	63	32	-10	23S	149W
TX	Clipperton Is	NA	10	07	-7	10N	109W
TX	Marquesas Is	Oc	63	31	-10	9S	140W
TX	St Pierre & Miquelon	NA	09	05	-4	47N	56W
TX	Crozet	Af	68	39	+3	46S	52E
TX	Kerguelen Is	Af	68	39	+5	50S	70E
TX	Amsterdam & St Paul Is	Af	68	39	+5	38S	78E
TX	Wallis & Futuna Is	Oc	62	32	-10.5	14S	172W
TY	Benin	Af	46	35	+0	6N	3E
TZ	Mali	Af	46	35	+0	13N	8W
UA	Russia	Eu	19,20,29,30	16	+3	56N	37E
UA2	Kaliningrad	Eu	29	15	+2	55N	21E
UA8-0	Russia (Asiatic)	As	20-26,30-35,75	16-19,23	+7	52N	104E
UB	Ukraine	Eu	29	16	+2	50N	30E
UC	Belarus	Eu	29	16	+2	54N	28E
UD	Azerbaijan	As	29	21	+4	40N	50E
UF	Georgia	As	29	21	+4	42N	45E
UG	Armenia	As	29	21	+4	40N	45E
UI	Uzbekistan	As	30	17	+6	41N	69E
UJ	Tajikistan	As	30	17	+6	39N	69E
UK	Uzbekistan	As	30	17	+6	41N	69E

Prefix	Country Name	Continent	ITU Zone	CQ Zone	Time Zone vs UTC	Latitude	Longitude
UL	Kazakhstan	As	29-31	17	+5.5	43N	77E
UN	Kazakhstan	As	29-31	17	+5.5	43N	77E
UN1*	Karelo-Finnish Rep	Eu	19	16	+3	64N	32E
UO	Moldova	Eu	29	16	+3	47N	29E
UP	Lithuania	Eu	29	15	+2	55N	25E
UQ	Latvia	Eu	29	15	+2	57N	24E
UR	Ukraine	Eu	29	16	+2	50N	30E
UR	Estonia	Eu	29	15	+2	59N	25E
V2	Antigua, Barbuda	NA	11	08	-4	17N	62W
V3	Belize	NA	11	07	-5.5	17N	89W
V4	St Kitts, Nevis	NA	11	08	-4	17N	63W
V5	Namibia	Af	57	38	+2	22S	17E
V59A	PDR Yemen	As	39	21	+3	13N	45E
V59P	PDR Yemen	As	39	21	+3	13N	45E
V6	Micronesia	Oc	65	27	+11	7N	158E
V7	Marshall Is	Oc	65	31	+12	7N	171E
V8	Brunei	Oc	54	28	+8	5N	115E
VA	Canada	NA	02-04,09,75	01-05	-5	45N	76W
VE	Canada	NA	02-04,09,75	01-05	-5	45N	76W
VK	Australia	Oc	55,58,59	29,30	+10	35S	149E
VK0	Heard Is	Af	68	39	+5	53S	73E
VK0	Macquarie Is	Oc	60	30	+11	54S	159E
VK0	Antarctica	An	67,69-74	12,13,29,30,32,38,39	+0	90S	0W
VK9	Christmas Is	Oc	54	29	+7	10S	106E
VK9	Cocos-Keeling Is	Oc	54	29	+6.5	12S	97E
VK9	Lord Howe Is	Oc	60	30	+10	31S	159E
VK9	Mellish Reef	Oc	55	30	+10	17S	156E
VK9	Norfolk Is	Oc	60	32	+11.5	29S	168E
VK9	Willis Is	Oc	55	30	+10	16S	150E
VK9	Papua Terr	Oc	51	28	+10	10S	147E
VK9	Terr New Guinea	Oc	51	28	+10	10S	147E
VK9X	Christmas Is	Oc	54	29	+7	10S	106E
VO	Canada	NA	02-04,09,75	01-05	-5	45N	76W

Prefix	Country Name	Continent	ITU Zone	CQ Zone	Time Zone vs UTC	Latitude	Longitude
VO*	Newfoundland, Labrador	NA	09	02,05	-3.5	48N	53W
VP2E	Anguilla	NA	11	08	-4	18N	63W
VP2M	Montserrat	NA	11	08	-4	17N	62W
VP2V	Br Virgin Is	NA	11	08	-4	18N	65W
VP5	Turks & Caicos Is	NA	11	08	-5	22N	71W
VP6	Pitcairn Is	Oc	63	32	-8.5	25S	128W
VP6	Ducie Is	Oc	63	32	-8.5	25S	125W
VP8	Falkland Is	SA	16	13	-4	52S	58W
VP8	So Georgia Is	SA	73	13	-1.5	54S	37W
VP8	So Orkney Is	SA	73	13	-3	61S	45W
VP8	So Sandwich Is	SA	73	13	-3	59S	27W
VP8	So Shetland Is	SA	73	13	-4	62S	58W
VP8	Antarctica	An	67,69-74	12,13,29,30,32,38,39	+0	90S	0W
VP9	Bermuda	NA	11	05	-4	32N	65W
VQ1*	Zanzibar	Af	53	37	+3	7S	39E
VQ6*	British Somaliland	Af	48	37	+3	2N	46E
VQ9	Chagos	Af	41	39	+5	7S	72E
VQ9*	Aldabra	Af	53	39	+4	9S	46E
VQ9*	Desroches	Af	53	39	+4	6S	55E
VQ9*	Farquhar	Af	53	39	+4	10S	51E
VR	China	As	33,42-44	23,24	+8	40N	116E
VR2	Hong Kong	As	44	24	+8	22N	114E
VR6	Pitcairn Is	Oc	63	32	-8.5	25S	128W
VS2*	Malaya	As	54	28	+7.5	3N	102E
VS4*	Sarawak	Oc	54	28	+8	2N	110E
VS6	Hong Kong	As	44	24	+8	22N	114E
VS9H*	Kuria Muria Is	As	39	21	+4	18N	56E
VS9K*	Kamaran Is	As	39	21	+3	15N	43E
VS9S	PDR Yemen	As	39	21	+3	13N	45E
VU	India	As	41	22	+5.5	29N	77E
VU4	Andaman & Nicobar Is	As	49	26	+5.5	12N	93E
VU7	Lakshadweep Is	As	41	22	+5.5	11N	73E
VY	Canada	NA	02-04,09,75	01-05	-5	45N	76W

Prefix	Country Name	Continent	ITU Zone	CQ Zone	Time Zone vs UTC	Latitude	Longitude
WH0	Mariana Is	Oc	64	27	+10	15N	146E
WH1	Baker & Howland Is	Oc	61	31	-12	0N	176W
WH2	Guam	Oc	64	27	+10	13N	145E
WH3	Johnston Is	Oc	61	31	-11	17N	170W
WH4	Midway Is	Oc	61	31	-11	28N	177W
WH5	Palmyra, Jarvis Is	Oc	61,62	31	-11	6N	162W
WH5K	Kingman Reef	Oc	61	31	-11	6N	162W
WH6	Hawaii	Oc	61	31	-10	21N	158W
WH7K	Kure Is	Oc	61	31	-11	29N	178W
WH8	Am Samoa	Oc	62	32	-11	14S	171W
WH8S	Swain's Island	Oc	62	32	-11	11S	171W
WH9	Wake Is	Oc	65	31	+12	19N	167E
WHY	Hawaii	Oc	61	31	-10	21N	158W
WL	Alaska	NA	01,02	01	-8	58N	134W
WP1	Navassa Is	NA	11	08	-5	18N	75W
WP2	Virgin Is	NA	11	08	-4	18N	65W
WP3	Puerto Rico	NA	11	08	-4	18N	66W
WP4	Puerto Rico	NA	11	08	-4	18N	66W
WP5	Desecheo Is	NA	11	08	-4	18N	68W
XE	Mexico	NA	10	06	-6	20N	99W
XF4	Revilla Gigedo	NA	10	06	-7	18N	113W
XP	Greenland	NA	05,75	40	-3	64N	52W
XS	China	As	33,42-44	23,24	+8	40N	116E
XT	Burkina Faso	Af	46	35	+0	12N	2W
XU	Cambodia	As	49	26	+8	12N	105E
XV9	Spratly Is	As					
XW	Laos	As	49	26	+7	20N	102E
XX9	Macao	As	44	24	+8	22N	114E
XZ	Myanmar (Burma)	As	49	26	+6.5	17N	96E
Y2*	East Germany	Eu	28	14	+1	53N	13E
YB	Indonesia	Oc	51,54	28	+7.5	6S	107E
YI	Iraq	As	39	21	+3	32N	45E
YJ	Vanuatu	Oc	56	32	+11	18S	168E
YK	Syria	As	39	20	+2	34N	36E
YL	Latvia	Eu	29	15	+2	57N	24E

Prefix	Country Name	Continent	ITU Zone	CQ Zone	Time Zone vs UTC	Latitude	Longitude
YN	Nicaragua	NA	11	07	-6	12N	87W
YO	Romania	Eu	28	20	+2	45N	26E
YS	El Salvador	NA	11	07	-6	14N	89W
YU	Serbia	Eu	28	15	+1	45N	21E
YU3	Montenegro	Eu	28	15	+1	42N	19E
YU6	Montenegro	Eu	28	15	+1	42N	19E
YV	Venezuela	SA	12	09	-4	10N	67W
YV0	Aves Is	NA	11	08	-4	16N	64W
Z2	Zimbabwe	Af	53	38	+2	18S	31E
Z3	Macedonia	Eu	28	15	+1	42N	22E
Z8	South Sudan	Af	48	34	+2	5N	32E
ZA	Albania	Eu	28	15	+1	41N	20E
ZB	Gibraltar	Eu	37	14	+1	37N	5W
ZC	Cyprus SBA	As	39	20	+2	35N	33E
ZC5*	Br No Borneo	Oc	54	28	+8	6N	116E
ZC6*	Palestine	As	39	20	+2	32N	35E
ZD4*	Gold Coast, Togoland	Af	46	35	+0	5N	0W
ZD7	St Helena	Af	66	36	+0	16S	6W
ZD8	Ascension Is	Af	66	36	+0	8S	14W
ZD9	Tristan da Cunha & Gough Is	Af	66	38	+0	37S	12W
ZF	Cayman Is	NA	11	08	-5	19N	81W
ZK1	So Cook Is	Oc	63	32	-10.5	22S	158W
ZK1	No Cook Is	Oc	62,63	32	-10.5	10S	161W
ZK2	Niue	Oc	62	32	-11	19S	170W
ZK3	Tokelau Is	Oc	62	31	-11	9S	171W
ZL	New Zealand	Oc	60	32	+12	41S	175E
ZL5	Antarctica	An	67,69-74	12,13,29,30,32,38,39	+0	90S	0W
ZL7	Chatham Is	Oc	60	32	+12.75	44S	177W
ZL8	Kermadec Is	Oc	60	32	+12	29S	178W
ZL9	New Zealand Subantarctic Is	Oc	60	32	+12	51S	166E
ZP	Paraguay	SA	14	11	-4	26S	57W
ZS	So Africa	Af	57	38	+2	26S	28E
ZS0*	Penguin Is	Af	57	38	+2	27S	15E
ZS7	Antarctica	An	67,69-74	12,13,29,30,32,38,39	+0	90S	0W
ZS8	Pr Edward &	Af	57	38	+3	47S	38E

Prefix	Country Name	Continent	ITU Zone	CQ Zone	Time Zone vs UTC	Latitude	Longitude
	Marion Is						
ZS9*	Walvis Bay	Af	57	38	+2	23S	15E
ZX0	Antarctica	An	67,69-74	12,13,29,30,32,38,39	+0	90S	0W

* = Deleted country/entity. ~ = "Unofficial." Ø = "zero."

Data compilation by Bill Brelsford, K2DI. Some geographic information by Ron McConnell, W2IOL. revised data may be available from time to time at http://codxc.org.

World Call Sign Prefixes by Allocation

Country/DX Entity	Primary Amateur Prefix	Continent	ITU Zone	CQ Zone	Time Zone vs UTC	Latitude	Longitude	ITU Prefix Allocations	Other Amateur Prefixes
Abu Ail, Jabal at Tair	A1*	As	39	21	+2	14N	43E		J2/A
Afghanistan	T6	As	40	21	+4.5	35N	69E	T6, YA	
Agalega & St Brandon	3B6	Af	53	39	+4	10S	57E		3B7
Aland Is	OHØ	Eu	18	15	+2	60N	20E		
Alaska	KL	NA	01, 02	01	-8	58N	134W		AL, NL, WL
Albania	ZA	Eu	28	15	+1	41N	20E	ZA	
Aldabra	VQ9*	Af	53	39	+4	9S	46E		
Algeria	7X	Af	37	33	+0	37N	3E	7R, 7T-7Y	
Am Samoa	KH8	Oc	62	32	-11	14S	171W		AH8, NH8, WH8
Amsterdam & St Paul Is	FT_Z	Af	68	39	+5	38S	78E		TX
Andaman & Nicobar Is	VU4	As	49	26	+5.5	12N	93E		
Andorra	C3	Eu	27	14	+1	43N	2E	C3	
Angola	D2	Af	52	36	+1	9S	13E	D2-D3	
Anguilla	VP2E	NA	11	08	-4	18N	63W		
Annobon Is	3CØ	Af	52	36	-1	1S	6E		
Antarctica	CE9	An	67, 69-74	12, 13, 29, 30, 32, 38, 39	+0	90S	0W		3Y, 8J1, ATØ, DPØ, FT_Y, KC4, LU_Z, OR4, RI1AN, VKØ, VP8, ZL5, ZS7, ZXØ
Antigua, Barbuda	V2	NA	11	08	-4	17N	62W	V2	
Argentina	LU	SA	14, 16	13	-3	35S	58W	AY-AZ, LO-LW, L2-L9	
Armenia	EK	As	29	21	+4	40N	45E	EK	UG
Aruba	P4	SA	11	09	-4	13N	70W	P4	
Ascension Is	ZD8	Af	66	36	+0	8S	14W		

Country/DX Entity	Primary Amateur Prefix	Continent	ITU Zone	CQ Zone	Time Zone vs UTC	Latitude	Longitude	ITU Prefix Allocations	Other Amateur Prefixes
Austral Is	FOØ	Oc	63	32	-10	23S	149W		TX
Australia	VK	Oc	55, 58, 59	29, 30	+10	35S	149E	AX, VH-VN, VZ	
Austria	OE	Eu	28	15	+1	48N	16E	OE	
Aves Is	YVØ	NA	11	08	-4	16N	64W		
Azerbaijan	4J	As	29	21	+4	40N	50E	4J-4K	UD
Azores	CU	Eu	36	14	-1	38N	26W		
Bahamas	C6	NA	11	08	-5	25N	77W	C6	
Bahrain	A9	As	39	21	+4	26N	51E	A9	
Bajo Nuevo	HKØ*	NA	11	08	-5	16N	79W		
Baker & Howland Is	KH1	Oc	61	31	-12	0N	176W		AH1, NH1, WH1
Balearic Is	EA6	Eu	37	14	+1	38N	3E		
Banaba	T33	Oc	65	31	+11.5	1S	170E		
Bangladesh	S2	As	41	22	+6	24N	90E	S2-S3	
Barbados	8P	NA	11	08	-4	13N	60W	8P	
Belarus	EV	Eu	29	16	+2	54N	28E	EU-EW	UC
Belgium	ON	Eu	27	14	+1	51N	4E	ON-OT	
Belize	V3	NA	11	07	-5.5	17N	89W	V3	
Benin	TY	Af	46	35	+0	6N	3E	TY	
Bermuda	VP9	NA	11	05	-4	32N	65W		
Bhutan	A5	As	41	22	+5.5	28N	90E	A5	
Blenheim Reef	~~*	Af	41	39	+5	7S	72E		
Bolivia	CP	SA	12, 14	10	-4	17S	68W	CP	
Bonaire	PJ4	SA	11	09	-4	12N	68W		
Bosnia-Hercegovina	E7	Eu	28	15	+1	44N	18E	E7	T9
Botswana	A2	Af	57	38	+2	25S	26E	8O, A2	
Bouvet	3Y	Af	67	38	+0	54S	3E		
Br No Borneo	ZC5*	Oc	54	28	+8	6N	116E		
Br Virgin Is	VP2V	NA	11	08	-4	18N	65W		
Brazil	PY	SA	12, 13, 15	11	-3	16S	48W	PP-PY, ZV-ZZ	
British Somaliland	VQ6*	Af	48	37	+3	2N	46E		
Brunei	V8	Oc	54	28	+8	5N	115E	V8	
Bulgaria	LZ	Eu	28	20	+2	43N	23E	LZ	
Burkina Faso	XT	Af	46	35	+0	12N	2W	XT	
Burundi	9U	Af	52	36	+3	3S	29E	9U	
Cambodia	XU	As	49	26	+8	12N	105E	XU	
Cameroon	TJ	Af	47	36	+1	4N	12E	TJ	

Country/DX Entity	Primary Amateur Prefix	Continent	ITU Zone	CQ Zone	Time Zone vs UTC	Latitude	Longitude	ITU Prefix Allocations	Other Amateur Prefixes
Canada	VE	NA	02-04, 09, 75	01-05	-5	45N	76W	CF-CK, CY-CZ, VA-VG, VO, VX-VY, XJ-XO	VA, VO, VY
Canal Zone	KZ5*	NA	11	07	-5	9N	80W		
Canary Is	EA8	Af	36	33	+0	28N	15W		
Cape Verde	D4	Af	46	35	-2	15N	23W	D4	
Cayman Is	ZF	NA	11	08	-5	19N	81W		
Celebe & Molucca Is	PK6*	Oc	54	28	+8	5S	119E		
Central African Rep	TL	Af	47	36	+1	5N	19E	TL	
Central Kiribati	T31	Oc	62	31	+12	4S	171W	T3	
Ceuta & Melilla	EA9	Af	37	33	+1	36N	5W		
Chad	TT	Af	47	36	+1	12N	15E	TT	
Chagos	VQ9	Af	41	39	+5	7S	72E		
Chatham Is	ZL7	Oc	60	32	+12.75	44S	177W		
Chesterfield Is	FK	Oc	55	30	+11	20S	158E		TX
Chile	CE	SA	14, 16	12	-4	33S	71W	3G, CA-CE, XQ-XR	
China	BY	As	33, 42-44	23, 24	+8	40N	116E	3H-3U, BA-BZ, VR, XS	
Christmas Is	VK9	Oc	54	29	+7	10S	106E		VK9X
Clipperton Is	FO0	NA	10	07	-7	10N	109W		TX
Cocos Is	TI9	NA	11	07	-6	6N	87W		
Cocos-Keeling Is	VK9	Oc	54	29	+6.5	12S	97E		VK9C
Colombia	HK	SA	12	09	-5	5N	74W	5J-5K, HJ-HK	
Comoros	FH*	Af	53	39	+3	12S	43E		FB8
Comoros	D6	Af	53	39	+3	12S	43E	D6	
Congo	TN	Af	52	36	+1	4S	15E	TN	
Conway Reef	3D2	Oc	56	32	+12	22S	175E		
Corsica	TK	Eu	28	15	+1	42N	9E		
Costa Rica	TI	NA	11	07	-6	10N	84W	TE, TI	
Crete	SV9	Eu	28	20	+2	36N	24E		
Croatia	9A	Eu	28	15	+1	46N	16E	9A	
Crozet	FT_W	Af	68	39	+3	46S	52E		TX
Cuba	CO	NA	11	08	-5	23N	82W	CL-CM, CO, T4	
Curacao	PJ2	SA	11	09	-4	12N	69W		
Cyprus	5B	As	39	20	+3	35N	33E	5B, C4, H2, P3	

Country/DX Entity	Primary Amateur Prefix	Continent	ITU Zone	CQ Zone	Time Zone vs UTC	Latitude	Longitude	ITU Prefix Allocations	Other Amateur Prefixes
Cyprus SBA	ZC	As	39	20	+2	35N	33E		
Czech Republic	OK	Eu	28	15	+1	50N	15E	OK-OL	
Czechoslovakia	OK*	Eu	28	15	+1	50N	15E		
Damao, Diu	CR8*	As	41	22	+5.5	21N	71E		
Dem Rep Congo (Zaire)	9Q	Af	52	36	+1	4S	15E	9O-9T	
Denmark	OZ	Eu	18	14	+1	56N	13E	5P-5Q, OU-OZ, XP	
Desecheo Is	KP5	NA	11	08	-4	18N	68W		NP5, WP5
Desroches	VQ9*	Af	53	39	+4	6S	55E		
Djibouti	J2	Af	48	37	+3	12N	43E	J2	
Dodecanese	SV5	Eu	28	20	+2	36N	28E		
Dominica	J7	NA	11	08	-4	15N	61W	J7	
Dominican Rep	HI	NA	11	08	-5	18N	70W	HI	
Ducie Is	VP6	Oc	63	32	-8.5	25S	125W		
East Germany	Y2*	Eu	28	14	+1	53N	13E		DM
East Kiribati	T32	Oc	61, 63	31	+12	2N	158W		
East Malaysia	9M6	Oc	54	28	+8	2N	110E		9M8
Easter Is	CE0Y	SA	63	12	-7	27S	109W		
Ecuador	HC	SA	12	10	-5	0N	79W	HC-HD	
Egypt	SU	Af, As	38	34	+2	31N	31E	6A-6B, SSA-SSM, SU	
El Salvador	YS	NA	11	07	-6	14N	89W	HU, YS	
England	G	Eu	27	14	+0	52N	0W	2A-2Z, GA-GZ, MA-MZ, VP-VQ, VS, ZB-ZJ, ZN-ZO, ZQ	2E, GX, M, MX
Equatorial Guinea	3C	Af	47	36	-1	4N	9E	3C	
Eritrea	E3	Af	48	37	+3	15N	39E	E3	ET2
Estonia	ES	Eu	29	15	+2	59N	25E	ES	UR
Ethiopia	ET	Af	48	37	+3	9N	39E	9E-9F, ET	
Falkland Is	VP8	SA	16	13	-4	52S	58W		
Faroe Is	OY	Eu	18	14	+0	62N	7W		OW
Farquhar	VQ9*	Af	53	39	+4	10S	51E		
Fernando de Noronha	PY0F	SA	13	11	-2	4S	32W		PY0ZF
Fiji	3D2	Oc	56	32	+12	18S	178E	3DN-3DZ	
Finland	OH	Eu	18	15	+2	60N	25E	OF-OJ	
Fr Equatorial Africa	FQ8*	Af	47, 52	36	+1	5N	18E		
Fr Indo China	FI8*	As	49	26	+7	11N	107E		

Country/DX Entity	Primary Amateur Prefix	Continent	ITU Zone	CQ Zone	Time Zone vs UTC	Latitude	Longitude	ITU Prefix Allocations	Other Amateur Prefixes
France	F	Eu	27	14	+1	49N	2E	FA-FZ, HW-HY, TH, TK, TM, TO-TQ, TV-TX	
Franz Josef Land	RI1F	Eu	75	40	+3	81N	48E		
French Guiana	FY	SA	12	09	-4	5N	52W		TO
French India	FN8*	As	41	22	+5.5	12N	80E		
French Polynesia	FO	Oc	63	32	-10	18S	150W		TX
French W Africa	FF*	Af	46	35	+0	15N	18W		
Gabon	TR	Af	52	36	+1	1N	10E	TR	
Galapagos Is	HC8	SA	12	10	-6	1S	90W		
Gambia	C5	Af	46	35	+0	13N	17W	C5	
Georgia	4L	As	29	21	+4	42N	45E	4L	UF
Germany	DL*	Eu	28	14	+1	52N	7E		
Germany	DL	Eu	28	14	+1	53N	13E	DA-DR, Y2-Y9	
Geyser Reef	~~*	Af	53	39	+3	12S	46E		
Ghana	9G	Af	46	35	+0	5N	0W	9G	
Gibraltar	ZB	Eu	37	14	+1	37N	5W		
Glorioso Is	FT_G	Af	53	39	+3	12S	47E		FR/G, TO
Goa	CR8*	As	41	22	+5.5	16N	74E		
Gold Coast, Togoland	ZD4*	Af	46	35	+0	5N	0W		
Greece	SV	Eu	28	20	+2	38N	24E	J4, SV-SZ	
Greenland	OX	NA	05, 75	40	-3	64N	52W		XP
Grenada	J3	NA	11	08	-4	12N	62W	J3	
Guadeloupe	FG	NA	11	08	-4	16N	62W		TO
Guam	KH2	Oc	64	27	+10	13N	145E		AH2, NH2, WH2
Guantanamo Bay	KG4	NA	11	08	-5	20N	75W		
Guatemala	TG	NA	11	07	-6	16N	92W	TD, TG	
Guernsey	GU	Eu	27	14	+0	49N	3W		2U, GP, MU, MP
Guinea	3X	Af	46	35	+0	10N	14W	3X	
Guinea-Bissau	J5	Af	46	35	-1	12N	16W	J5	
Guyana	8R	SA	12	09	-3.75	6N	58W	8R	
Haiti	HH	NA	11	08	-5	19N	72W	4V, HH	
Hawaii	KH6	Oc	61	31	-10	21N	158W		AH6, NH6, WH6, AH7, KH7, NH7, WH7
Heard Is	VK0	Af	68	39	+5	53S	73E		
Honduras	HR	NA	11	07	-6	14N	87W	HQ-HR	
Hong Kong	VR2	As	44	24	+8	22N	114E		VS6

Country/DX Entity	Primary Amateur Prefix	Continent	ITU Zone	CQ Zone	Time Zone vs UTC	Latitude	Longitude	ITU Prefix Allocations	Other Amateur Prefixes
Hungary	HA	Eu	28	15	+1	48N	19E	HA, HG	
Iceland	TF	Eu	17	40	+0	64N	22W	TF	
Ifni	EA9*	Af	37	33	+0	29N	10W		
India	VU	As	41	22	+5.5	29N	77E	8T-8Y, AT-AW, VT-VW	
Indonesia	YB	Oc	51, 54	28	+7.5	6S	107E	7A-7I, 8A-8I, JZ, PK-PO, YB-YH	
International Civil Aviation Organization	4Y							4Y	
Iran	EP	As	40	21	+3.5	36N	51E	9B-9D, EP-EQ	
Iraq	YI	As	39	21	+3	32N	45E	HN, YI	
Ireland	EI	Eu	27	14	+0	53N	6W	EI-EJ	
Isle of Man	GD	Eu	27	14	+0	54N	4W		2D, GT, MD, MT
Israel	4X	As	39	20	+2	32N	35E	4X, 4Z	
Ital Somaliland	I5*	Af	48	37	+3	2N	46E		
Italy	I	Eu, Af	28, 37	15, 33	+1	42N	12E	IA-IZ	
ITU Geneva	4U	Eu	28	14	+1	46N	6E		
Ivory Coast	TU	Af	46	35	+0	7N	5W	TU	
Jamaica	6Y	NA	11	08	-5	18N	77W	6Y	
Jan Mayen	JX	Eu	18	40	-1	71N	9W		
Japan	JA	As	45	25	+9	36N	140E	7J-7N, 8J-8N, JA-JS	
Java	PK1*	Oc	54	28	+7.5	6S	107E		PK2-3
Jersey	GJ	Eu	27	14	+0	49N	2W		2J, GH, MJ, MH
Johnston Is	KH3	Oc	61	31	-11	17N	170W		AH3, NH3, WH3
Jordan	JY	As	39	20	+2	32N	36E	JY	
Juan de Nova, Europa	FT_J	Af	53	39	+3	17S	43E		FT_E, FR/J, TO
Juan Fernandez	CE0Z	SA	14	12	-4	34S	79W		
Kaliningrad	RA2	Eu	29	15	+2	55N	21E		UA2
Kamaran Is	VS9K*	As	39	21	+3	15N	43E		7O
Karelo-Finnish Rep	UN1*	Eu	19	16	+3	64N	32E		
Kazakhstan	UN	As	29-31	17	+5.5	43N	77E	UN-UQ	UL
Kenya	5Z	Af	48	37	+3	2S	37E	5Y-5Z	
Kerguelen Is	FT_X	Af	68	39	+5	50S	70E		TX
Kermadec Is	ZL8	Oc	60	32	+12	29S	178W		
Kingman Reef	KH5K*	Oc	61	31	-11	6N	162W		AH5K, NH5K, WH5K
Kosovo	Z6	Eu	28	15	+1	43N	21E	Z6~	

Country/DX Entity	Primary Amateur Prefix	Continent	ITU Zone	CQ Zone	Time Zone vs UTC	Latitude	Longitude	ITU Prefix Allocations	Other Amateur Prefixes
Kure Is	KH7K	Oc	61	31	-11	29N	178W		AH7K, NH7K, WH7K
Kuria Muria Is	VS9H*	As	39	21	+4	18N	56E		
Kuwait	9K	As	39	21	+3	29N	48E	9K	
Kuwait/S Arabia NZ	8Z5*	As	39	21	+3	29N	48E		9K3
Kyrgyzstan	EX	As	30, 31	17	+6	43N	75E	EX	UM
Lakshadweep Is	VU7	As	41	22	+5.5	11N	73E		
Laos	XW	As	49	26	+7	20N	102E	XW	
Latvia	YL	Eu	29	15	+2	57N	24E	YL	UQ
Lebanon	OD	As	39	20	+2	34N	36E	OD	
Lesotho	7P	Af	57	38	+2	29S	27E	7P	
Liberia	EL	Af	46	35	-0.75	6N	11W	5L-5M, 6Z, A8, D5, EL	
Libya	5A	Af	38	34	+2	33N	13E	5A	
Liechtenstein	HBØ	Eu	28	14	+1	47N	10E		
Lithuania	LY	Eu	29	15	+2	55N	25E	LY	UP
Lord Howe Is	VK9	Oc	60	30	+10	31S	159E		VK9L
Luxembourg	LX	Eu	27	14	+1	50N	6E	LX	
Macao	XX9	As	44	24	+8	22N	114E		
Macedonia	Z3	Eu	28	15	+1	42N	22E	Z3	4N5
Macquarie Is	VKØ	Oc	60	30	+11	54S	159E		
Madagascar	5R	Af	53	39	+3	19S	48E	5R-5S, 6X	
Madeira Is	CT3	Af	36	33	-1	33N	17W		
Malawi	7Q	Af	53	37	+2	14S	34E	7Q	
Malaya	VS2*	As	54	28	+7.5	3N	102E		9M2
Maldives	8Q	As, Af	41	22	+5	4N	73E	8Q	
Mali	TZ	Af	46	35	+0	13N	8W	TZ	
Malpelo Is	HKØ	SA	12	09	-5	4N	82W		
Malta	9H	Eu	28	15	+1	36N	15E	9H	
Malyj Vysotskij Is	RI1M*	Eu	29	16	+3	61N	29E		
Manchuria	C9*	As	33	24	+8.5	46N	127E		
Mariana Is	KHØ	Oc	64	27	+10	15N	146E		AHØ, NHØ, WHØ
Market Reef	OJØ	Eu	18	15	+2	60N	19E		
Marquesas Is	FOØ	Oc	63	31	-10	9S	140W		TX
Marshall Is	V7	Oc	65	31	+12	7N	171E	V7	KX6
Martinique	FM	NA	11	08	-4	15N	61W		TO
Mauritania	5T	Af	46	35	-1	18N	16W	5T	
Mauritius	3B8	Af	53	39	+4	20S	58E	3B	

Country/DX Entity	Primary Amateur Prefix	Continent	ITU Zone	CQ Zone	Time Zone vs UTC	Latitude	Longitude	ITU Prefix Allocations	Other Amateur Prefixes
Mayotte	FH	Af	53	39	+3	13S	45E		TX
Mellish Reef	VK9	Oc	55	30	+10	17S	156E		VK9M
Mexico	XE	NA	10	06	-6	20N	99W	4A-4C, 6D-6J, XA-XI	
Micronesia	V6	Oc	65	27	+11	7N	158E	V6	KC6
Midway Is	KH4	Oc	61	31	-11	28N	177W		AH4, NH4, WH4
Minami Torishima	JD	Oc	90	27	+10	24N	154E		
Minerva Reef	1M*	Oc	62	32	-12	24S	179W		
Moldova	ER	Eu	29	16	+3	47N	29E	ER	UO
Monaco	3A	Eu	27	14	+1	44N	8E	3A	
Mongolia	JT	As	32, 33	23	+7.5	48N	107E	JT-JV	
Montenegro	4O	Eu	28	15	+1	42N	19E	4O	YU3, YU6,..
Montserrat	VP2M	NA	11	08	-4	17N	62W		
Morocco	CN	Af	37	33	+0	34N	7W	5C-5G, CN	
Mozambique	C9	Af	53	37	+2	26S	33E	C8-C9	
Mt Athos	SV/A	Eu	28	20	+2	40N	24E		SY2
Myanmar (Burma)	XZ	As	49	26	+6.5	17N	96E	XY-XZ	
Namibia	V5	Af	57	38	+2	22S	17E	V5	
Nauru	C2	Oc	65	31	+11.5	1S	167E	C2	
Navassa Is	KP1	NA	11	08	-5	18N	75W		NP1, WP1
Nepal	9N	As	42	22	+5.75	28N	85E	9N	
Neth Antilles	PJ2*	SA	11	09	-4	12N	69W		PJ4, PJ9
Neth New Guinea	JZ0*	Oc	51	28	+10	10S	147E		
Netherlands	PA	Eu	27	14	+1	52N	5E	PA-PJ	
Netherlands Borneo	PK5*	Oc	54	28	+8	3S	115E		
New Caledonia	FK	Oc	56	32	+11	22S	167E		TX
New Zealand	ZL	Oc	60	32	+12	41S	175E	ZK-ZM	
New Zealand Subantarctic Is	ZL9	Oc	60	32	+12	51S	166E		
Newfoundland, Labrador	VO*	NA	09	02, 05	-3.5	48N	53W		
Nicaragua	YN	NA	11	07	-6	12N	87W	HT, H6-H7, YN	
Niger	5U	Af	46	35	+1	14N	2W	5U	
Nigeria	5N	Af	46	35	+1	6N	3E	5N-5O	
Niue	E6	Oc	62	32	-11	19S	170W	E6	ZK2
No Cook Is	E5	Oc	62, 63	32	-10.5	10S	161W		ZK1
No Ireland	GI	Eu	27	14	+0	55N	6W		2I, GN, MI, MN
No Korea	P5	As	44	25	+9	39N	126E	HM, P5-P9	
Norfolk Is	VK9	Oc	60	32	+11.5	29S	168E		VK9N

Country/DX Entity	Primary Amateur Prefix	Continent	ITU Zone	CQ Zone	Time Zone vs UTC	Latitude	Longitude	ITU Prefix Allocations	Other Amateur Prefixes
Norway	LA	Eu	18	14	+1	60N	11E	3Y, JW-JX, LA-LN	
Ogasawara	JD	As	45	27	+10	28N	142E		
Okinawa (Ryukyu)	KR6*	As	45	25	+8	26N	128E		KR8, JR6, KA6
Okino Tori-shima	JD1*	As	45	27	+10	30N	140E		7J1
Oman	A4	As	39	21	+4	24N	59E	A4	
Pakistan	AP	As	41	21	+5	34N	73E	6P-6S, AP-AS	
Palau	T8	Oc	64	27	+10	7N	134E	T8	KC6
Palestine	E4	As	39	20	+2	32N	34E	E4	
Palestine	ZC6*	As	39	20	+2	32N	35E		4X1
Palmyra, Jarvis Is	KH5	Oc	61, 62	31	-11	6N	162W		AH5, NH5, WH5
Panama	HP	NA	11	07	-5	9N	80W	3E-3F, HO-HP, H3, H8-H9	
Papua New Guinea	P2	Oc	51	28	+10	10S	147E	P2	
Papua Terr	P2*	Oc	51	28	+10	10S	147E		VK9
Paraguay	ZP	SA	14	11	-4	26S	57W	ZP	
PDR Yemen	7O*	As	39	21	+3	13N	45E		VS9A, VS9P, VS9S
Penguin Is	ZSØ*	Af	57	38	+2	27S	15E		
Peru	OA	SA	12	10	-5	12S	78W	4T, OA-OC	
Peter I	3Y	An	72	12	-6	69S	91W		
Philippines	DU	Oc	50	27	+8	15N	121E	4D-4I, DU-DZ	
Pitcairn Is	VP6	Oc	63	32	-8.5	25S	128W		VR6
Poland	SP	Eu	28	15	+1	52N	21E	3Z, HF, SN-SR	
Portugal	CT	Eu	37	14	+0	39N	9W	CQ-CU, XX	
Portuguese Timor	CR8*	Oc	54	28	+8	9S	126E		
Pr Edward & Marion Is	ZS8	Af	57	38	+3	47S	38E		
Pratas Is	BV9	As	44	24	+8	21N	116E		BQ9
Puerto Rico	KP4	NA	11	08	-4	18N	66W		KP3, NP3, WP3, NP4, WP4
Qatar	A7	As	39	21	+4	25N	52E	A7	
Reunion	FR	Af	53	39	+4	21S	55E		TO
Revilla Gigedo	XF4	NA	10	06	-7	18N	113W		
Rodriguez Is	3B9	Af	53	39	+4	20S	63E		
Romania	YO	Eu	28	20	+2	45N	26E	YO-YR	
Rotuma	3D2	Oc	56	32	+12	13S	177E		

Country/DX Entity	Primary Amateur Prefix	Continent	ITU Zone	CQ Zone	Time Zone vs UTC	Latitude	Longitude	ITU Prefix Allocations	Other Amateur Prefixes
Ruanda-Urundi	9U5*	Af	52	36	+3	3S	30E		
Russia	R	Eu	19, 20, 29, 30	16	+3	56N	37E	RA-RZ, UA-UI	UA
Russia (Asiatic)	R8	As	20-26, 30-35, 75	16-19, 23	+7	52N	104E		R9-Ø, UA8-Ø
Rwanda	9X	Af	52	36	+3	2S	30E	9X	
S Arabia/Iraq NZ	8Z4*	As	39	21	+3	29N	46E		
Saar	9S4*	Eu	28	14	+1	49N	7E		
Saba & St Eustatius	PJ5	NA	11	08	-4	18N	63W		PJ6
Sable Is	CYØ	NA	09	05	-5	44N	60W		
Samoa	5W	Oc	62	32	-11	14S	172W	5W	
San Andres & Providencia	HKØ	NA	11	07	-6	13N	82W		
San Felix	CEØX	SA	14	12	-5	26S	80W		
San Marino	T7	Eu	28	15	+1	44N	12E	T7	
Sao Tome & Principe	S9	Af	47	36	+0	0N	7E	S9	
Sarawak	VS4*	Oc	54	28	+8	2N	110E		
Sardinia	IS	Eu	28	15	+1	39N	9E		
Saudi Arabia	HZ	As	39	21	+3	25N	47E	8Z, HZ, 7Z	
Scarborough Reef	BS7	As	50	27	+8	15N	118E		
Scotland	GM	Eu	27	14	+0	57N	2W		2M, GS, GZ, MM, MS, MZ
Senegal	6W	Af	46	35	+0	15N	18W	6V-6W	
Serbia	YU	Eu	28	15	+1	45N	21E	YT-YU	
Serrana Bnk, Roncador Cay	HKØ*	NA	11	07	-5	14N	80W		KP3, KS4
Seychelles	S7	Af	53	39	+4	5S	55E	S7	
Sierra Leone	9L	Af	46	35	+0	9N	13W	9L	
Sikkim	AC3*	As	41	22	+5.5	27N	89E		
Singapore	9V	As	54	28	+7.5	1N	104E	9V, S6	
Slovakia	OM	Eu	28	15	+1	48N	17E	OM	
Slovenia	S5	Eu	28	15	+1	46N	15E	S5	
SMO Malta	1A	Eu	28	15	+1	42N	13E	1A~	
So Africa	ZS	Af	57	38	+2	26S	28E	H5~, S4~, S8~, V9~, ZR-ZU	
So Cook Is	E5	Oc	63	32	-10.5	22S	158W	E5	ZK1
So Georgia Is	VP8	SA	73	13	-1.5	54S	37W		LU_Z
So Korea	HL	As	44	25	+9	38N	127E	6K-6N, DS-DT, D7-D9, HL	
So Orkney Is	VP8	SA	73	13	-3	61S	45W		LU_Z

Country/DX Entity	Primary Amateur Prefix	Continent	ITU Zone	CQ Zone	Time Zone vs UTC	Latitude	Longitude	ITU Prefix Allocations	Other Amateur Prefixes
So Sandwich Is	VP8	SA	73	13	-3	59S	27W		LU_Z
So Shetland Is	VP8	SA	73	13	-4	62S	58W		LU_Z, CE9, HFØ, RI1AN
Solomon Is	H4	Oc	51	28	+11	9S	160E	H4	
Somalia	T5	Af	48	37	+3	2N	46E	6O, T5	
South Sudan	Z8	Af	48	34	+2	5N	32E	Z8	
Southern Sudan	STØ*	Af	48	34	+2	5N	32E		
Spain	EA	Eu	37	14	+1	40N	4W	AM-AO, EA-EH	
Spratly Is	1S	As	50	26	+7	9N	112E	1S~	9MØ, XV9
Sri Lanka	4S	As	41	22	+5.5	7N	80E	4P-4S	
St Barthelemy	FJ	NA	11	08	-4	18N	63W		TO
St Helena	ZD7	Af	66	36	+0	16S	6W		
St Kitts, Nevis	V4	NA	11	08	-4	17N	63W	V4	
St Lucia	J6	NA	11	08	-4	14N	61W	J6	
St Maarten	PJ7	NA	11	08	-4	18N	63W		
St Maarten, Saba, St Eus	PJ5*	NA	11	08	-4	18N	63W		PJ6-8
St Martin	FS	NA	11	08	-4	18N	63W		TO
St Paul Is	CY9	NA	09	05	-5	47N	60W		
St Peter & St Paul Rocks	PYØP	SA	13	11	-2	1N	29W		PYØZP
St Pierre & Miquelon	FP	NA	09	05	-4	47N	56W		TX
St Vincent	J8	NA	11	08	-4	13N	61W	J8	
Sudan	ST	Af	48	34	+2	16N	33E	6T-6U, SSN-SSZ, ST	
Sumatra	PK4*	Oc	54	28	+7	1S	100E		
Surinam	PZ	SA	12	09	-3.5	6N	55W	PZ	
Svalbard	JW	Eu	18	40	+1	78N	16E		
Swain's Island	KH8S	Oc	62	32	-11	11S	171W		AH8S, NH8S, WH8S
Swan Is	KS4*	NA	11	07	-6	17N	84W		
Swaziland	3DA	Af	57	38	+2	26S	31E	3DA-3DM	3D6
Sweden	SM	Eu	18	14	+1	59N	18E	7S, 8S, SA-SM	
Switzerland	HB	Eu	28	14	+1	47N	7E	HB, HE	
Syria	YK	As	39	20	+2	34N	36E	6C, YK	
Taiwan	BV	As	44	24	+8	25N	122E		BM-BQ, BU, BW, BX
Tajikistan	EY	As	30	17	+6	39N	69E	EY	UJ
Tangier	CN2*	Af	37	33	+0	36N	8W		
Tanzania	5H	Af	53	37	+3	7S	39E	5H-5I	

Country/DX Entity	Primary Amateur Prefix	Continent	ITU Zone	CQ Zone	Time Zone vs UTC	Latitude	Longitude	ITU Prefix Allocations	Other Amateur Prefixes
Temotu	H4Ø	Oc	51	28	+11	11S	166E		
Terr New Guinea	P2*	Oc	51	28	+10	10S	147E		VK9
Thailand	HS	As	49	26	+7	14N	101E	E2, HS	
Tibet	AC4*	As	41	23	+6	30N	92E		
Timor Leste	4W	Oc	54	28	+8	9S	126E	4W	
Togo	5V	Af	46	35	+0	6N	1E	5V	
Tokelau Is	ZK3	Oc	62	31	-11	9S	171W		
Tonga	A3	Oc	62	32	+13	21S	175W	A3	
Trieste	I1*	Eu	28	15	+1	46N	14E		
Trindade & Martin Vaz Is	PYØT	SA	15	11	-2	21S	29W		PYØZT
Trinidad & Tobago	9Y	SA	11	09	-4	11N	62W	9Y-9Z	
Tristan da Cunha & Gough Is	ZD9	Af	66	38	+0	37S	12W		
Tromelin	FT_T	Af	53	39	+4	16S	54E		FR/T, TO
Tunisia	3V	Af	37	33	+1	37N	10E	3V, TS	
Turkey	TA	As, Eu	39	20	+2	40N	33E	TA-TC, YM	
Turkmenistan	EZ	As	30	17	+5	38N	58E	EZ	UH
Turks & Caicos Is	VP5	NA	11	08	-5	22N	71W		
Tuvalu	T2	Oc	65	31	+12	9S	179E	T2	
Uganda	5X	Af	48	37	+3	0N	33E	5X	
Ukraine	UR	Eu	29	16	+2	50N	30E	EM-EO, UR-UZ	UB
UN HQ	4U	NA	08	05	-5	41N	74W	4U	
United Arab Emirates	A6	As	39	21	+4	24N	54E	A6	
United States	K	NA	06-08	03-05	-5	39N	77W	AA-AL, KA-KZ, NA-NZ, WA-WZ	
Uruguay	CX	SA	14	13	-3	35S	56W	CV-CX	
Uzbekistan	UK*	As	30	17	+6	41N	69E	UJ-UM	UI
Vanuatu	YJ	Oc	56	32	+11	18S	168E	YJ	
Vatican	HV	Eu	28	15	+1	42N	13E	HV	
Venezuela	YV	SA	12	09	-4	10N	67W	4M, YV-YY	
Vietnam	3W	As	49	26	+7	11N	107E	3W, XV	
Virgin Is	KP2	NA	11	08	-4	18N	65W		NP2, WP2
Wake Is	KH9	Oc	65	31	+12	19N	167E		AH9, NH9, WH9
Wales	GW	Eu	27	14	+0	52N	3W		2W, GC, MW, MC
Wallis & Futuna Is	FW	Oc	62	32	-10.5	14S	172W		TX
Walvis Bay	ZS9*	Af	57	38	+2	23S	15E		
West Kiribati	T3Ø	Oc	65	31	+12	1S	173E		

Country/DX Entity	Primary Amateur Prefix	Continent	ITU Zone	CQ Zone	Time Zone vs UTC	Latitude	Longitude	ITU Prefix Allocations	Other Amateur Prefixes
West Malaysia	9M2	As	54	28	+7.5	3N	102E	9M, 9W	9M4
Western Sahara	SØ	Af	46	33	+0	27N	13W	SØ~	
Willis Is	VK9	Oc	55	30	+10	16S	150E		VK9W
World Meteorological Organization	C7							C7	
Yemen	7O	As	39	21	+3	13N	45E	7O	
Yemen Arab Rep	4W*	As	39	21	+3	15N	44E		
Zambia	9J	Af	53	36	+2	15S	28E	9I-9J	
Zanzibar	VQ1*	Af	53	37	+3	7S	39E		5H1
Zimbabwe	Z2	Af	53	38	+2	18S	31E	Z2	

* = Deleted country/entity. ~ = "Unofficial." Ø = "zero."

Data compilation by Bill Brelsford, K2DI. Some geographic information by Ron McConnell, W2IOL. revised data may be available from time to time at http://codxc.org.

Third Party Operating Agreements

Call Sign Prefix	Country or Entity	Call Sign Prefix	Country or Entity
V2	Antigua/Barbuda	6Y	Jamaica
LO-LW	Argentina	JY	Jordan
VK	Australia	EL	Liberia
V3	Belize	V7	Marshall Islands
CP	Bolivia	XA-XI	Mexico
E7	Bosnia-Herzegovina	V6	Micronesia, Federated States of
PP-PY	Brazil	YN	Nicaragua
VE, VO, VY	Canada	HO-HP	Panama
CA-CE	Chile	ZP	Paraguay
HJ-HK	Colombia	OA-OC	Peru
D6	Comoros (Federal Islamic Republic of)	DU-DZ	Philippines
TI, TE	Costa Rica	VR6	Pitcairn Island*
CM, CO	Cuba	V4	St. Kitts/Nevis
HI	Dominican Republic	J6	St. Lucia
J7	Dominica	J8	St. Vincent and the Grenadines
HC-HD	Ecuador	9L	Sierra Leone
YS	El Salvador	ZR-ZU	South Africa
C5	Gambia, The	3DA	Swaziland
9G	Ghana	9Y-9Z	Trinidad/Tobago
J3	Grenada	TA-TC	Turkey
TG	Guatemala	GB	United Kingdom
8R	Guyana	CV-CX	Uruguay
HH	Haiti	YV-YY	Venezuela
HQ-HR	Honduras	4U1ITU	ITU - Geneva
4X, 4Z	Israel	4U1VIC	VIC - Vienna

* Since 1970, there has been an informal agreement between the United Kingdom and the US, permitting Pitcairn and US amateurs to exchange messages concerning medical emergencies, urgent need for equipment or supplies, and private or personal matters of island residents.

US licensed amateurs may operate in the US territories under their FCC license. Note: At the end of an exchange of third-party traffic with a station located in a foreign country, an FCC-licensed amateur must transmit the call sign of the foreign station as well as his/her own call sign.

World Time Chart

UTC	0000	0100	0200	0300	0400	0500	0600	0700	0800	0900	1000	1100	1200	1300	1400	1500	1600	1700	1800	1900	2000	2100	2200	2300
Central Europe, Stockholm, Vienna	1AM	2AM	3AM	4AM	5AM	6AM	7AM	8AM	9AM	10AM	11AM	12N	1PM	2PM	3PM	4PM	5PM	6PM	7PM	8PM`	9PM	10PM	11PM	12M
Eastern Europe, Moscow, Cape Town	2AM	3AM	4AM	5AM	6AM	7AM	8AM	9AM	10AM	11AM	12N	1PM	2PM	3PM	4PM	5PM	6PM	7PM	8PM`	9PM	10PM	11PM	12M	1AM
Saudi Arabia	3AM	4AM	5AM	6AM	7AM	8AM	9AM	10AM	11AM	12N	1PM	2PM	3PM	4PM	5PM	6PM	7PM	8PM`	9PM	10PM	11PM	12M	1AM	2AM
Iran	4AM	5AM	6AM	7AM	8AM	9AM	10AM	11AM	12N	1PM	2PM	3PM	4PM	5PM	6PM	7PM	8PM`	9PM	10PM	11PM	12M	1AM	2AM	3AM
Central Russia, Mumbai	5AM	6AM	7AM	8AM	9AM	10AM	11AM	12N	1PM	2PM	3PM	4PM	5PM	6PM	7PM	8PM`	9PM	10PM	11PM	12M	1AM	2AM	3AM	4AM
Tibet, Kolkata	6AM	7AM	8AM	9AM	10AM	11AM	12N	1PM	2PM	3PM	4PM	5PM	6PM	7PM	8PM`	9PM	10PM	11PM	12M	1AM	2AM	3AM	4AM	5AM
China, Taiwan, Hong Kong	7AM	8AM	9AM	10AM	11AM	12N	1PM	2PM	3PM	4PM	5PM	6PM	7PM	8PM`	9PM	10PM	11PM	12M	1AM	2AM	3AM	4AM	5AM	6AM
Philippines, Perth	8AM	9AM	10AM	11AM	12N	1PM	2PM	3PM	4PM	5PM	6PM	7PM	8PM`	9PM	10PM	11PM	12M	1AM	2AM	3AM	4AM	5AM	6AM	7AM
Tokyo	9AM	10AM	11AM	12N	1PM	2PM	3PM	4PM	5PM	6PM	7PM	8PM`	9PM	10PM	11PM	12M	1AM	2AM	3AM	4AM	5AM	6AM	7AM	8AM
Melbourne	10AM	11AM	12N	1PM	2PM	3PM	4PM	5PM	6PM	7PM	8PM`	9PM	10PM	11PM	12M	1AM	2AM	3AM	4AM	5AM	6AM	7AM	8AM	9AM
New Zealand	11AM	12N	1PM	2PM	3PM	4PM	5PM	6PM	7PM	8PM`	9PM	10PM	11PM	12M	1AM	2AM	3AM	4AM	5AM	6AM	7AM	8AM	9AM	10AM
International Date Line	12N	1PM	2PM	3PM	4PM	5PM	6PM	7PM	8PM`	9PM	10PM	11PM	12M	1AM	2AM	3AM	4AM	5AM	6AM	7AM	8AM	9AM	10AM	11AM
Nome, Alaska	1PM	2PM	3PM	4PM	5PM	6PM	7PM	8PM`	9PM	10PM	11PM	12M	1AM	2AM	3AM	4AM	5AM	6AM	7AM	8AM	9AM	10AM	11AM	12N
Hawaii	2PM	3PM	4PM	5PM	6PM	7PM	8PM`	9PM	10PM	11PM	12M	1AM	2AM	3AM	4AM	5AM	6AM	7AM	8AM	9AM	10AM	11AM	12N	1PM
Eastern Alaska	3PM	4PM	5PM	6PM	7PM	8PM`	9PM	10PM	11PM	12M	1AM	2AM	3AM	4AM	5AM	6AM	7AM	8AM	9AM	10AM	11AM	12N	1PM	2PM
Vancouver, BC, Seattle, Las Vegas (PST)	4PM	5PM	6PM	7PM	8PM`	9PM	10PM	11PM	12M	1AM	2AM	3AM	4AM	5AM	6AM	7AM	8AM	9AM	10AM	11AM	12N	1PM	2PM	3PM
Boise, Salt Lake City, Denver, Calgary {MST}	5PM	6PM	7PM	8PM`	9PM	10PM	11PM	12M	1AM	2AM	3AM	4AM	5AM	6AM	7AM	8AM	9AM	10AM	11AM	12N	1PM	2PM	3PM	4PM
Oklahoma City, Chicago, Kansas City, Winnipeg (CST)	6PM	7PM	8PM`	9PM	10PM	11PM	12M	1AM	2AM	3AM	4AM	5AM	6AM	7AM	8AM	9AM	10AM	11AM	12N	1PM	2PM	3PM	4PM	5PM
Indianapolis, Montreal, New York City (EST)	7PM	8PM`	9PM	10PM	11PM	12M	1AM	2AM	3AM	4AM	5AM	6AM	7AM	8AM	9AM	10AM	11AM	12N	1PM	2PM	3PM	4PM	5PM	6PM
Nova Scotia, Venezuela, Rio de Janeiro (AST)	8PM`	9PM	10PM	11PM	12M	1AM	2AM	3AM	4AM	5AM	6AM	7AM	8AM	9AM	10AM	11AM	12N	1PM	2PM	3PM	4PM	5PM	6PM	7PM
Greenland	9PM	10PM	11PM	12M	1AM	2AM	3AM	4AM	5AM	6AM	7AM	8AM	9AM	10AM	11AM	12N	1PM	2PM	3PM	4PM	5PM	6PM	7PM	8PM
Azores	10PM	11PM	12M	1AM	2AM	3AM	4AM	5AM	6AM	7AM	8AM	9AM	10AM	11AM	12N	1PM	2PM	3PM	4PM	5PM	6PM	7PM	8PM	9PM
Cabo Verde	11PM	12M	1AM	2AM	3AM	4AM	5AM	6AM	7AM	8AM	9AM	10AM	11AM	12N	1PM	2PM	3PM	4PM	5PM	6PM	7PM	8PM	9PM	10PM

Note: Iran, Afghanistan, India, Myanmar, and central Australia are offset 30 minutes. Chart does not account for Daylight Savings Time.

World Time by Country

Country	No. of time zones	Time zone
Afghanistan	1	UTC+04:30
Albania	1	UTC+01:00 (CET)
Algeria	1	UTC+01:00 (CET)
Andorra	1	UTC+01:00 (CET)
Angola	1	UTC+01:00 (WAT)
Antarctica	9	UTC−03:00 (ART) — Palmer Station, Rothera Station UTC±00:00 (GMT) — Troll Station UTC+03:00 (UTC+03:00) — Syowa Station UTC+05:00 — Mawson Station UTC+06:00 — Vostok Station UTC+07:00 (UTC+07:00) — Davis Station UTC+10:00 — Dumont-d'Urville Station UTC+11:00 — Casey Station, Macquarie Island UTC+12:00 — McMurdo Station
Antigua and Barbuda	1	UTC−04:00 (AST)
Argentina	1	UTC−03:00 (ART)
Armenia	1	UTC+04:00
Australia	9	UTC+05:00 — Heard and McDonald Islands UTC+06:30 — Cocos (Keeling) Islands UTC+07:00 (CXT) — Christmas Island UTC+08:00 (AWST) — Western Australia, Indian Pacific railroad when travelling between Port Augusta, South Australia and Kalgoorlie, Western Australia) UTC+08:45 (CWT) – South Australia (Border Village), Western Australia (Caiguna, Eucla, Madura, Mundrabilla) UTC+09:30 (ACST) — South Australia, Northern Territory, New South Wales (Yancowinna County) UTC+10:00 (AEST) — Queensland, New South Wales, Australian Capital Territory, Victoria, Tasmania UTC+10:30 — Lord Howe Island UTC+11:00 (NFT) — Norfolk Island
Austria	1	UTC+01:00 (CET)
Azerbaijan	1	UTC+04:00
Bahamas	1	UTC−05:00 (EST)
Bahrain	1	UTC+03:00
Bangladesh	1	UTC+06:00 (BDT)
Barbados	1	UTC−04:00
Belarus	1	UTC+03:00 (FET)

Country	No. of time zones	Time zone
Belgium	1	UTC+01:00 (CET)
Belize	1	UTC−06:00
Benin	1	UTC+01:00 (WAT)
Bhutan	1	UTC+06:00 (BTT)
Bolivia	1	UTC−04:00
Bosnia and Herzegovina	1	UTC+01:00 (CET)
Botswana	1	UTC+02:00 (CAT)
Brazil	4	UTC−05:00 (Brasília time −2) — Acre and Southwestern Amazonas UTC−04:00 (Brasília time −1) — Most part of the Amazonas State, Mato Grosso, Mato Grosso do Sul, Rondônia, Roraima UTC−03:00 (Brasília time) — the Southeast Region, the South Region, the Northeast Region (except some islands), Goiás, Distrito Federal, Tocantins, Pará, Amapá UTC−02:00 (Brasília time +1) — few islands on the east coast of Brazil (Fernando de Noronha, Trindade and Martim Vaz, Rocas Atoll, Saint Peter and Saint Paul Archipelago)
Brunei	1	UTC+08:00
Bulgaria	1	UTC+02:00 (EET)
Burkina Faso	1	UTC±00:00
Burundi	1	UTC+02:00 (CAT)
Cambodia	1	UTC+07:00
Cameroon	1	UTC+01:00 (WAT)
Canada	6	UTC−08:00 (PT) — larger western part of British Columbia, Tungsten and the associated Cantung Mine in Northwest Territories, Yukon UTC−07:00 (MT) — Alberta, some eastern parts of British Columbia, most of Northwest Territories, Nunavut (west of 102°W and all communities in the Kitikmeot Region), Lloydminster and surrounding area in Saskatchewan UTC−06:00 (CT) — Manitoba, Nunavut (between 85° West and 102°W except western Southampton Island), Ontario (Northwestern Ontario west of 90°W with some exceptions and Big Trout Lake area east of 90°W), Saskatchewan except Lloydminster UTC−05:00 (ET) — Nunavut east of 85°W and entire Southampton Island, Ontario east of 90°W (except Big Trout Lake area) plus several more western areas, Quebec (most of province) UTC−04:00 (AT) — Labrador (all but southeastern tip), New Brunswick, Nova Scotia, Prince Edward Island, eastern part of Quebec UTC−03:30 (NT) — Labrador (southeastern), Newfoundland
Cape Verde	1	UTC−01:00 (Cape Verde Time)
Central African Republic	1	UTC+01:00 (WAT)
Chad	1	UTC+01:00 (WAT)

Country	No. of time zones	Time zone
Chile	3	UTC−06:00 — Easter Island UTC−04:00 — main territory UTC−03:00 — Magallanes and Chilean Antarctica
China	2	UTC+06:00 (Xinjiang Time) UTC+08:00 (Chinese Standard Time)
China	1	UTC+08:00 (Chinese Standard Time)
Colombia	1	UTC−05:00
Comoros	1	UTC+03:00 (EAT)
Costa Rica	1	UTC−06:00
Croatia	1	UTC+01:00 (CET)
Cuba	1	UTC−05:00
Cyprus	1	UTC+02:00 (EET)
Czech Republic	1	UTC+01:00 (CET) (CRT)
Democratic Republic of the Congo	2	UTC+01:00 (WAT) — western part of the country UTC+02:00 (CAT) — eastern part of the country
Denmark	5	UTC−04:00 — Thule Air Base in Greenland UTC−03:00 — most of Greenland, including inhabited south coast and west coast UTC−01:00 — Ittoqqortoormiit and surrounding area in Greenland's Tunu county UTC±00:00 — Danmarkshavn weather station and surrounding area in Greenland's Tunu county, Faroe Islands UTC+01:00 — (CET) — metropolitan Denmark
Djibouti	1	UTC+03:00 (EAT)
Dominica	1	UTC−04:00
Dominican Republic	1	UTC−04:00
East Timor	1	UTC+09:00
Ecuador	2	UTC−06:00 (GALT) — Galápagos Province UTC−05:00 (Ecuador Time) — main territory of Ecuador
Egypt	1	UTC+02:00 (EET)
El Salvador	1	UTC−06:00
Equatorial Guinea	1	UTC+01:00 (WAT)
Eritrea	1	UTC+03:00 (EAT)
Estonia	1	UTC+02:00 (EET)
Eswatini (Swaziland)	1	UTC+02:00

Country	No. of time zones	Time zone
Ethiopia	1	UTC+03:00 (EAT)
Federated States of Micronesia	2	UTC+10:00 — the states of Chuuk and Yap UTC+11:00 — the states of Kosrae and Pohnpei
Fiji	1	UTC+12:00
Finland	1	UTC+02:00 (EET)
France	12	UTC−10:00 — most of French Polynesia UTC−09:30 — Marquesas Islands UTC−09:00 — Gambier Islands UTC−08:00 — Clipperton Island UTC−04:00 (AST) — Guadeloupe, Martinique, Saint Barthélemy, Saint Martin UTC−03:00 (PMST) — French Guiana, Saint Pierre and Miquelon UTC+01:00 (CET) — Metropolitan France UTC+03:00 — Mayotte UTC+04:00 — Réunion, Crozet Islands, Scattered Islands in the Indian Ocean UTC+05:00 — Kerguelen Islands, Îles Saint-Paul and Amsterdam UTC+11:00 — New Caledonia UTC+12:00 — Wallis and Futuna
Gabon	1	UTC+01:00 (WAT)
Gambia	1	UTC±00:00
Georgia	1	UTC+04:00
Germany	1	UTC+01:00 (CET)
Ghana	1	UTC±00:00
Greece	1	UTC+02:00 (EET)
Grenada	1	UTC−04:00
Guatemala	1	UTC−06:00
Guinea	1	UTC±00:00
Guinea-Bissau	1	UTC±00:00
Guyana	1	UTC−04:00
Haiti	1	UTC−05:00
Honduras	1	UTC−06:00
Hong Kong (China)	1	UTC+08:00 (HKT)
Hungary	1	UTC+01:00 (CET)
Iceland	1	UTC±00:00
India	1	UTC+05:30 (IST)

Country	No. of time zones	Time zone
Indonesia	3	UTC+07:00 (Western Indonesian Standard Time) — islands of Sumatra, Java, provinces of West Kalimantan and Central Kalimantan UTC+08:00 (Central Indonesian Standard Time) — islands of Sulawesi, Bali, provinces of East Nusa Tenggara, West Nusa Tenggara, East Kalimantan, North Kalimantan and South Kalimantan UTC+09:00 (Eastern Indonesian Standard Time) — provinces of Maluku, North Maluku, Papua and West Papua
Iran	1	UTC+03:30 (IRST)
Iraq	1	UTC+03:00
Ireland	1	UTC±00:00 (WET)
Israel	1	UTC+02:00 (IST)
Italy	1	UTC+01:00 (CET)
Ivory Coast	1	UTC±00:00
Jamaica	1	UTC−05:00
Japan	1	UTC+09:00 (JST)
Jordan	1	UTC+02:00
Kazakhstan	2	UTC+05:00 — western Kazakhstan UTC+06:00 — eastern Kazakhstan
Kenya	1	UTC+03:00 (EAT)
Kingdom of the Netherlands	2	UTC−04:00 (AST) — Caribbean municipalities and constituent countries UTC+01:00 (CET) — main territory of the Netherlands
Kiribati	3	UTC+12:00 — Gilbert Islands UTC+13:00 — Phoenix Islands UTC+14:00 — Line Islands
Korea, North	1	UTC+09:00 (Pyongyang Time)
Korea, South	1	UTC+09:00 (Korea Standard Time)
Kosovo	1	UTC+01:00
Kuwait	1	UTC+03:00 (Arabia Standard Time)
Kyrgyzstan	1	UTC+06:00
Laos	1	UTC+07:00
Latvia	1	UTC+02:00 (EET)
Lebanon	1	UTC+02:00 (EET)
Lesotho	1	UTC+02:00
Liberia	1	UTC±00:00

Country	No. of time zones	Time zone
Libya	1	UTC+02:00 (EET)
Liechtenstein	1	UTC+01:00 (CET)
Lithuania	1	UTC+02:00 (EET)
Luxembourg	1	UTC+01:00 (CET)
Macau (China)	1	UTC+08:00 (Macau Standard Time)
Madagascar	1	UTC+03:00 (EAT)
Malawi	1	UTC+02:00 (CAT)
Malaysia	1	UTC+08:00 (Malaysian Standard Time)
Maldives	1	UTC+05:00
Mali	1	UTC±00:00
Malta	1	UTC+01:00 (CET)
Marshall Islands	1	UTC+12:00
Mauritania	1	UTC±00:00
Mauritius	1	UTC+04:00 (Mauritius Time)
Mexico	4	UTC−08:00 (Zone 4 or Northwest Zone) — the state of Baja California UTC−07:00 (Zone 3 or Pacific Zone) — the states of Baja California Sur, Chihuahua, Nayarit, Sinaloa and Sonora UTC−06:00 (Zone 2 or Central Zone) — most of Mexico UTC−05:00 (Zone 1 or Southeast Zone) — the state of Quintana Roo
Moldova	1	UTC+02:00 (EET)
Monaco	1	UTC+01:00 (CET)
Mongolia	2	UTC+07:00 — the provinces of Khovd, Uvs and Bayan-Ölgii UTC+08:00 — most of the country
Montenegro	1	UTC+01:00 (CET)
Morocco	1	UTC+01:00 (CET)[5]
Mozambique	1	UTC+02:00 (CAT)
Myanmar	1	UTC+06:30 (MST)
Namibia	1	UTC+01:00 (WAT)
Nauru	1	UTC+12:00
Nepal	1	UTC+05:45 (Nepal Time)

Country	No. of time zones	Time zone
New Zealand	5	UTC−11:00 — Niue UTC−10:00 — Cook Islands UTC+12:00 — main territory of New Zealand UTC+12:45 — Chatham Islands UTC+13:00 — Tokelau
Nicaragua	1	UTC−06:00
Niger	1	UTC+01:00 (WAT)
Nigeria	1	UTC+01:00 (WAT)
North Macedonia	1	UTC+01:00 (CET)
Norway	1	UTC+01:00 (CET)
Oman	1	UTC+04:00
Pakistan	1	UTC+05:00 (PKT)
Palau	1	UTC+09:00
Palestine	1	UTC+02:00
Panama	1	UTC−05:00
Papua New Guinea	2	UTC+10:00 — most of the country UTC+11:00 — Autonomous Region of Bougainville (Bougainville Standard Time)
Paraguay	1	UTC−04:00
Peru	1	UTC−05:00 (PET)
Philippines	1	UTC+08:00 (PHT)
Poland	1	UTC+01:00 (CET)
Portugal	2	UTC−01:00 — Azores UTC±00:00 (WET) — Madeira and the main territory of Portugal
Qatar	1	UTC+03:00 (Arabia Standard Time)
Republic of the Congo	1	UTC+01:00 (WAT)
Romania	1	UTC+02:00 (EET)

Country	No. of time zones	Time zone
Russia	11	UTC+02:00 (Kaliningrad Time) — Kaliningrad Oblast UTC+03:00 (Moscow Time) — Most of European Russia and all railroads throughout Russia UTC+04:00 (Samara Time) — Astrakhan Oblast, Samara Oblast, Saratov Oblast, Udmurtia, and Ulyanovsk Oblast UTC+05:00 (Yekaterinburg Time) — Bashkortostan, Chelyabinsk Oblast, Khanty-Mansia, Kurgan Oblast, Orenburg Oblast, Perm Krai, Sverdlovsk Oblast, Tyumen Oblast, and Yamalia UTC+06:00 (Omsk Time) — Omsk Oblast UTC+07:00 (Krasnoyarsk Time) — Altai Krai, Altai Republic, Kemerovo Oblast, Khakassia, Krasnoyarsk Krai, Novosibirsk Oblast, Tomsk Oblast, and Tuva UTC+08:00 (Irkutsk Time) — Buryatia and Irkutsk Oblast UTC+09:00 (Yakutsk Time) — Amur Oblast, western Sakha Republic, and Zabaykalsky Krai UTC+10:00 (Vladivostok Time) — Jewish Autonomous Oblast, Khabarovsk Krai, Primorsky Krai, and central Sakha Republic UTC+11:00 (Magadan Time) — Magadan Oblast, eastern Sakha, and Sakhalin Oblast UTC+12:00 (Kamchatka Time) — Chukotka and Kamchatka Krai
Rwanda	1	UTC+02:00 (CAT)
Saint Kitts and Nevis	1	UTC−04:00
Saint Lucia	1	UTC−04:00
Saint Vincent and the Grenadines	1	UTC−04:00
Samoa	1	UTC+13:00
San Marino	1	UTC+01:00 (CET)
São Tomé and Príncipe	1	UTC+00:00
Saudi Arabia	1	UTC+03:00 (Arabia Standard Time)
Senegal	1	UTC±00:00
Serbia	1	UTC+01:00 (CET)
Seychelles	1	UTC+04:00 (Seychelles Time)
Sierra Leone	1	UTC±00:00
Singapore	1	UTC+08:00 (SST)
Slovakia	1	UTC+01:00 (CET)
Slovenia	1	UTC+01:00 (CET)
Solomon Islands	1	UTC+11:00
Somalia	1	UTC+03:00 (EAT)
South Africa	2	UTC+02:00 (South African Standard Time) — main territory UTC+03:00 — Prince Edward Islands
South Sudan	1	UTC+03:00 (EAT)

Country	No. of time zones	Time zone
Spain	2	UTC±00:00 (WET) — Canary Islands UTC+01:00 (CET) — main territory of Spain
Sri Lanka	1	UTC+05:30 (SLST)
Sudan	1	UTC+02:00
Suriname	1	UTC−03:00
Sweden	1	UTC+01:00 (CET)
Switzerland	1	UTC+01:00 (CET)
Syria	1	UTC+02:00 (EET)
Taiwan	1	UTC+08:00
Tajikistan	1	UTC+05:00
Tanzania	1	UTC+03:00 (EAT)
Thailand	1	UTC+07:00 (THA)
Togo	1	UTC±00:00
Tonga	1	UTC+13:00
Trinidad and Tobago	1	UTC−04:00 (AST)
Tunisia	1	UTC+01:00 (CET)
Turkey	1	UTC+03:00 (TRT)
Turkmenistan	1	UTC+05:00
Tuvalu	1	UTC+12:00
Uganda	1	UTC+03:00 (EAT)
Ukraine	2	UTC+02:00[1] (EET) — most of the country UTC+03:00 — part of Donetsk and Luhansk regions[2][3][4]
United Arab Emirates	1	UTC+04:00
United Kingdom	9	UTC−08:00 — Pitcairn Islands UTC−05:00 — Cayman Islands, Turks and Caicos Islands UTC−04:00 (AST) — Anguilla, Bermuda, British Virgin Islands, Montserrat UTC−03:00 (FKST) — Falkland Islands UTC−02:00 — South Georgia and the South Sandwich Islands UTC±00:00 (GMT in winter/BST in summer) — main territory of the United Kingdom, Saint Helena, Ascension and Tristan da Cunha, Guernsey, Isle of Man, Jersey UTC+01:00 (CET) — Gibraltar UTC+02:00 (EET) — Akrotiri and Dhekelia UTC+06:00 — British Indian Ocean Territory

Country	No. of time zones	Time zone
United States	11	UTC−12:00 (unofficial) — Baker Island and Howland Island UTC−11:00 (ST) — American Samoa, Jarvis Island, Kingman Reef, Midway Atoll and Palmyra Atoll UTC−10:00 (HT) — Hawaii, most of the Aleutian Islands, and Johnston Atoll UTC−09:00 (AKT) — most of the state of Alaska UTC−08:00 (PT) — Pacific Time zone: the Pacific coast states and Nevada UTC−07:00 (MT) — Mountain Time zone: the Mountain states plus western parts of some adjacent states UTC−06:00 (CT) — Central Time zone: a large area spanning from the Gulf Coast to the Great Lakes UTC−05:00 (ET) — Eastern Time zone: roughly a triangle covering all the states from the Great Lakes down to Florida and east to the Atlantic coast UTC−04:00 (AST) — Puerto Rico, the U.S. Virgin Islands UTC+10:00 (ChT) — Guam and the Northern Mariana Islands UTC+12:00 (unofficial) — Wake Island, McMurdo Station, and Amundsen–Scott South Pole Station
Uruguay	1	UTC−03:00
Uzbekistan	1	UTC+05:00 (Uzbekistan Time)
Vanuatu	1	UTC+11:00
Vatican City	1	UTC+01:00 (CET)
Venezuela	1	UTC−04:00
Vietnam	1	UTC+07:00 (Indochina Time)
Yemen	1	UTC+03:00
Zambia	1	UTC+02:00 (CAT)
Zimbabwe	1	UTC+02:00 (CAT)

RST Signal Report Guidelines

	Readability
1	Unreadable. "I can tell there's a signal there, but I have no idea what you are sending."
2	Barely readable. Occasional words (or letters for CW) are getting through.
3	Readable with considerable difficulty.
4	Readable with almost no difficulty.
5	Perfectly readable. "Sounds like you're right next door."
	Signal Strength
1	Faint signal; barely perceptible.
2	Very weak signal.
3	Weak signal.
4	Fair signal.
5	Fairly good signal.
6	Good signal.
7	Moderately strong signal.
8	Strong signal.
9	Extremely strong signal.
	Tone (Applies Only to CW)
1	60 Hz AC or less, very rough and broad.
2	Very rough AC; very harsh and broad.
3	Rough AC tone, rectified but not filtered.
4	Rough note, some trace of filtering.
5	Filtered, rectified AC but strongly ripple modulated.
6	Filtered tone, definite trace of ripple modulation.
7	Near pure tone, trace of ripple modulation.
8	Near perfect tone, slight trace of modulation.
9	Perfect tone, no trace of ripple or modulation of any kind.
	Additional Descriptors for CW
X	Tone sounds very steady, like crystal (XTAL) control.
C	Tone is chirpy.
K	Tone is clicky.

Q Signals

The most used Q signs are in **boldface**.

Sign	Code	Question	Answer or Notice
QRA	— · — · — · — ·	What is the name (or call sign) of your station?	The name (or call sign) of my station is ...
QRG	— — · — · — — ·	Will you tell me my exact frequency (or that of ...)?	Your exact frequency (or that of ...) is ... kHz (or MHz).
QRH	— — · — · · · ·	Does my frequency vary?	Your frequency varies.
QRI	— — · — · ·	How is the tone of my transmission?	The tone of your transmission is (1. Good; 2. Variable; 3. Bad)
QRJ	— — · — · — — —	How many voice contacts do you want to make?	I want to make ... voice contacts.
QRK	— — · — — · —	What is the readability of my signals (or those of ...)?	The readability of your signals (or those of ...) is ... (1 to 5).
QRL	— — · — · — · ·	Are you busy? Is this frequency being used?	I am busy. (or I am busy with ...) Please do not interfere.
QRM	— — · — — —	Do you have interference? [from other stations]	I have interference.
QRN	— — · — — ·	Are you troubled by static?	I am troubled by static.
QRO	— — · — — — —	Shall I increase power?	Increase power.
QRP	— — · — · — — ·	Shall I decrease power?	Decrease power.
QRQ	— — · — — — · —	Shall I send faster?	Send faster (... wpm)
QRS	— — · — · · ·	Shall I send more slowly?	Send more slowly (... wpm)
QRT	— — · — —	Shall I cease or suspend operation?/ shutoff the radio	I am suspending operation. /shutting off the radio
QRU	— — · — · · —	Have you anything for me?	I have nothing for you.
QRV	— — · — · · · —	Are you ready?	I am ready.
QRW	— — · — · — —	Shall I inform ... that you are calling him on ... kHz (or MHz)?	Please inform ... that I am calling him on ... kHz (or MHz).
QRX	— — · — — · · —	Shall I standby / When will you call me again?	Please standby / I will call you again at ... (hours) on ... kHz (or MHz)
QRZ	— — · — — — · ·	Who is calling me?	You are being called by ... on ... kHz (or MHz)
QSA	— — · · · —	What is the strength of my signals (or those of ...)?	The strength of your signals (or those of ...) is ... (1 to 5).
QSB	— — · · · · ·	Are my signals fading?	Your signals are fading.
QSD	— — · · — · ·	Is my keying defective?	Your keying is defective.

Sign	Code	Question	Answer or Notice
QSG	— — · — · · · — — ·	Shall I send ... telegrams (messages) at a time?	Send ... telegrams (messages) at a time.
QSK	— — · — · · · — · —	Can you hear me between your signals? (Are you in full break-in mode?)	I can hear you between my signals. (I am in full break-in mode.)
QSL	— — · — · · · · — · ·	Can you acknowledge receipt?	I am acknowledging receipt.
QSM	— — · — · · · — —	Shall I repeat the last telegram (message) which I sent you, or some previous telegram (message)?	Repeat the last telegram (message) which you sent me.
QSN	— — · — · · · — ·	Did you hear me (or ... (call sign)) on .. kHz (or MHz)?	I did hear you (or ... (call sign)) on ... kHz (or MHz).
QSO	— — · — · · · — — —	Can you communicate with ... direct or by relay?	I can communicate with ... direct (or by relay through _____). Less formally, a contact or conversation.
QSP	— — · — · · · · — — ·	Will you relay a message to ...?	I will relay a message to ....
QSR	— — · — · · · · — ·	Do you want me to repeat my call?	Please repeat your call; I did not hear you.
QSS	— — · — · · · · · ·	What working frequency will you use?	I will use the working frequency ... kHz (or MHz).
QST	— — · — · · · —		Here is a broadcast message to all amateurs.
QSU	— — · — · · · · · —	Shall I send or reply on this frequency (or on ... kHz (or MHz))?	Send or reply on this frequency (or on ... kHz (or MHz).
QSW	— — · — · · · · — —	Will you send on this frequency (or on ... kHz (or MHz))?	I am going to send on this frequency (or on ... kHz (or MHz)).
QSX	— — · — · · · — · · —	Will you listen to _____ (call sign(s) on _____ kHz (or MHz))?	I am listening to ... (call sign(s) on ... kHz (or MHz))
QSY	— — · — · · · — · — —	Shall I change to transmission on another frequency?	Change to transmission on another frequency (or on ... kHz (or MHz)).
QSZ	— — · — · · · — — · ·	Shall I send each word or group more than once?	Send each word or group twice (or ... times).
QTA	— — · — — · —	Shall I cancel telegram (message) No. ... as if it had not been sent?	Cancel telegram (message) No. ... as if it had not been sent.
QTC	— — · — — — · — ·	How many telegrams (messages) have you to send?	I have ... telegrams (messages) for you (or for ...).
QTH	— — · — — · · · ·	What is your location?	My location is ...
QTR	— — · — — · — ·	What is the correct time?	The correct time is ... hours
QTU	— — · — — · · —	At what times are you operating?	I am operating from ... to ... hours.

Sign	Code	Question	Answer or Notice
QTX	— — · — — — · · —	Will you keep your station open for further communication with me until further notice (or until ... hours)?	I will keep my station open for further communication with you until further notice (or until ... hours).
QUA	— — · — · · — · —	Have you news of ... (call sign)?	Here is news of ... (call sign).
QUC	— — · — · · — — · ·	What is the number (or other indication) of the last message you received from me (or from ... (call sign))?	The number (or other indication) of the last message I received from you (or from ... (call sign)) is ...
QUD	— — · — · · — — · ·	Have you received the urgency signal sent by ... (call sign of mobile station)?	I have received the urgency signal sent by ... (call sign of mobile station) at ... hours.
QUE	— — · — · · — ·	Can you speak in ... (language), – with interpreter if necessary; if so, on what frequencies?	I can speak in ... (language) on ... kHz (or MHz).
QUF	— — · — · · — · · — ·	Have you received the distress signal sent by ... (call sign of mobile station)?	I have received the distress signal sent by ... (call sign of mobile station) at ... hours.

Prosigns

Most common prosigns in **boldface**.

Prosign	Code	Phone Procedure Word	Explanation
73	— — · · · · · · — —	"Best regards/best wishes ..."	A friendly expression used at end of contact. Some say it is properly "7 - 3", not "seventy-three."
88	— — — · · — — · ·	"Love and kisses."	Save this one for your spouse or spousal equivalent!
AA	· — · —	All after	"The portion of the message to which I refer is all that follows the text _____."
AB	· — — · · ·	All before	"The portion of the message to which I refer is all that is before the text _____."
AR	· — · — ·	Out	**End of transmission / End of message**
AS	· — · · ·	Wait	"Am engaged in a contact with another station, please stand by." Or, simply, "I must pause for a few minutes."
AS AR	· — · · · · — · — ·	Wait, out	"I must pause for more than a few minutes."
BK	— · · · — ·	Break	Used to interrupt a transmission already in progress.
BN	— · · · — ·	All between	"The portion of the message to which I refer is all that falls between _____ and _____."
C	— · — ·	Correct/affirmative	Probably from Spanish, "si."
CFM	— · — · · — —	I acknowledge/confirm	Same as R.
CL	— · — · · — · ·	**Closing or clear.**	**Station is going off the air and will not listen to or answer any further calls. Sent after final identification.**
CP	— · — · · — — ·	Calling _____ or _____.	General call to two or more specific stations.

Prosign	Code	Phone Procedure Word	Explanation
CQ	—·—· —·—·	Calling any station	General call to all stations.
CS	—·—· ···	What is your call sign?	What is the name of your station?
DE	—·· ·	This is from	Used before the call sign of the calling station.
HH	········	Correction … (Oops!)	"Preceding text was in error. Corrected text follows." Sometimes written as $\overline{EEEEEEEE}$
$\overline{HH\ AR}$	········ ·—·—·	Disregard this transmission. Out.	"The entire message just sent is in error, disregard it."
HI HI	···· ·· ···· ··	Laughing – "LOL"	How to laugh in Morse Code.
K	—·—	Over	Used after calling CQ, or at the end of a transmission, to indicate any station is invited to transmit.
$\overline{KA}$	—·—·—	Attention	Message begins/start of work/new message
$\overline{KN}$		Over. Named station please respond.	"Go ahead, specific station."
N	—·	Negative	"Answer to question just received is 'no'."
NIL	—· ·· ·—··	Nothing heard	General-purpose response to any request or inquiry for which the answer is "nothing," "none," or "not available." Also used for "I have no messages for you."
OM	——— ——	"Old Man"	Friendly address to a male ham. "Hey, OM, how've you been?"
R	·—·	Roger	"Your last transmission was received." Does not indicate the message was understood, nor that it will be complied with.
$\overline{SK}$	···—·—	Closing/silent key.	Sent after final identification.
$\overline{SOS}$	···———···	Distress signal.	Only for imminent danger to life or property.
$\overline{VE}$	···—·	Verified	Message is verified.
WA	·—— ·—	Word after	
WB	·—— —···	Word before	
WX	·—— —··—	Weather	
YL	—·—— ·—··	Wife/girlfriend	Also used to identify women in general, as in, "We really need to get more YL's interested in ham radio."
ZWF	——·· ·—— ··—·	Wrong	Your last transmission was wrong. The correct version is _____.
?	··——··	Say again	Standing alone, a note of interrogation or request for repetition of a transmission not understood. When placed after a signal, modifies the signal to be a question or request.

ARRL QN Signals for CW Net Use

The QN signals listed below are special ARRL signals for use in amateur CW nets only. They are not for use in casual amateur conversation. Other meanings that may be used in other services do not apply. Do not use QN signals on phone nets. Say it with words. QN signals need not be followed by a question mark, even though the meaning may be interrogatory. QN signs marked with "*" are for Net Control station use only.

For Q-code assertions or queries which only need to be acknowledged as received, the usual practice is to respond with the letter "R" for "Roger" which means "Received correctly". Sending an "R" merely means the code has been correctly received and does not necessarily mean that the receiving operator has taken any other action. For Q-code queries that need to be answered in the affirmative, the usual practice is to respond with the letter "C" (Sounds like the Spanish word "Si"). For Q-code queries that need to be answered in the negative, the usual practice it to respond with the letter "N" for "no". For those Q-code assertions that merely need to be acknowledged as understood, the usual practice is to respond with the prosign SN or VE which means "understood".

Sign	Code	Question	Answer or Notice
QNA*	—— —·— ·—		Answer in prearranged order.
QNB*	—— —·— —· —···		Act as relay **B**etween _____ and _____
QNC	—— —·— —·—·		All net stations **C**opy. (I have a message for all net stations.)
QND*	—— —·— —··		Net is **D**irected (controlled by net control station).
QNE*	—— —·— ·		**E**ntire net stand by.
QNF	—— —·— ··—·		Net is **F**ree. (Not controlled.)
QNG	—— —·— ——·		_____ take over a net control station.
QNH	—— —·— ····		Your net frequency is **H**igh.
QNI	—— —·— ··	Net stations report **I**n.*	I am reporting **I**nto the net.
QNJ	—— —·— ·———	Can you copy me?	Can you copy _____?
QNK*	—— —·— —·—		Transmit message for _____ to _____.
QNL	—— —·— ·—··		Your net frequency is **L**ow.
QNM*	—— —·— ——		You are QRM'ing the net. Stand by.
QNN	—— —·— —·	What station has net control?	Net control station is _____.
QNO	—— —·— ———		Station is leaving the net.
QNP	—— —·— ·——·		Unable to copy you. Unable to copy _____.
QNQ*	—— —·— ——·—		Move frequency to _____ and wait for _____ to finish handling traffic. Then send him/her traffic for _____.
QNR	—— —·— ·—·		Answer _____ and **R**eceive traffic.
QNS	—— —·— ···	What **S**tations are in the net?	Following **S**tations are in the net.* (Follow with list.)
QNT	—— —·— —	I request permission to leave the net for _____ minutes.	
QNU*	—— —·— ··—		The net has traffic for you. Stand by.
QNV*	—— —·— ···—		Establish contact with _____ on this frequency. If successful, move to _____ and send him/her traffic for _____.
QNW	—— —·— ·——	How do I route messages for _____?	
QNX	—— —·— —··—	Request to be excused from the net.	You are excused from the net.*

Sign	Code	Question	Answer or Notice
QNY*	—— ·— ——·—		Shift to another frequency (or to _____ kHz) to clear traffic with _____.
QNZ	——·— ——· ——··		Zero beat your signal with mine.

ARRL Prosigns For CW Net Use

Prosign	Code	Phone Procedure Word	Explanation
$\overline{AA}$	·—·—	(Next line.)	Separation between parts of address or signature.
AA	·— ·—	All after	(Use to get fills.)
AB	·— —···	All before	(Use to get fills.)
ADEE	·— —·· · ·	Addressee	Name of person to whom message is addressed.
ADR	·— —·· ·—·	Address	(Second part of message.)
$\overline{AR}$	·—·—·	Out	End of transmission / End of message
ARL	·— ·—· ·—··	ARRL numbered message follows.	Indicates use of ARRL numbered message in text.
$\overline{AS}$	·—···	Stand by	Wait ...
B	—···	More	Another message to follow.
BK	—··· —·—	Break	Let me in.
$\overline{BT}$	—···—	Next section of message.	Indicates break between address and text, text and signature.
C	—·—·	Yes.	Correct, yes.
CFM	—·—· ··—· ——	Confirm, please.	Send back what I just sent.
CK	—·—· —·—	Check.	
DE	—·· ·	This is from _____	Used before the call sign of the calling station.
$\overline{HH}$	········	Correction.	"Preceding text was in error. Corrected text follows."
HX	···· —··—	Handling instructions.	Optional part of preamble. Single letters to follow.
HXA	···· —··— ·—	Handling instruction A	(followed by number) Collect landline delivery authorized by addressee within.... miles. (If no number, authorization is unlimited.)
HXB	···· —··— —···	Handling instruction B	(followed by number) Cancel message if not delivered within _____ hours of filing time; service originating station.
HXC	···· —··— —·—·	Handling instruction C	Report date and time of delivery (TOD) to originating station.

Prosign	Code	Phone Procedure Word	Explanation
HXD	···· —··— —··	Handling instruction D	Report to originating station the identity of station from which received, plus date and time. Report identity of station to which relayed, plus date and time, or if delivered report date, time and method of delivery.
HXE	···· —··— ·	Handling instruction E	Delivering station get reply from addressee, originate message back.
HXF	···· —··— ··—·	Handling instruction F	(Followed by number) Hold delivery until _____ (date).
HXG	···· —··— ——·	Handling instruction G	Delivery by mail or landline toll call not required. If toll or other expense involved, cancel message and service originating station.
IMI	··——··	Repeating ...	"I say again." Use for difficult or unusual words or groups.
K	—·—	Over	Invitation to transmit.
N	—·	No.	Negative, incorrect, or "no more messages to follow."
NR	—· ·—·	Number.	Message number _____ follows.
PBL	·—·· —··· ·—··	Preamble	First part of message.
R	·—·	Roger	Message received. Can also be used as a decimal point.
SIG	··· ·· ——·	Signature	Signed/signature (last section of message.)
SK	···—·—	Silent Key	Out, clear. End of communication, no reply expected.
TU	— ··—	Thank you.	
WA	·—— ·—	Word after.	(Used to get fills.)
WB	·—— —···	Word before.	(Used to get fills.)

ITU Emission Types

Part 97 refers to many different types of emissions, such as "A1C" or "J2D" but points you elsewhere -- §2.201 of the FCC Rules, *Emission, modulation and transmission characteristics,* for information on emission type designators. These designators were the creation of the International Telecommunications Union. Here's the code.

The full ITU designator begins with the bandwidth of the signal, in a form something like "6ØHØ", (60 Hz) "15ØH" (15Ø Hz) or "1M5Ø" (1.5 MHz.) Then comes the three to five character description of the signal.

The first letter describes the type of modulation of the emission.

Character	Description
N	Unmodulated carrier
A	Double-sideband amplitude modulation (e.g. AM broadcast radio)
B	Independent sideband (two sidebands containing different signals)
C	Vestigial sideband (e.g. NTSC)
D	Combination of AM and FM or PM
F	Frequency modulation (e.g. FM broadcast radio)
G	Phase modulation (can also be used to designate FM.)
H	Single-sideband with full carrier (e.g. as used by CHU, the Canadian version of WWV)
J	Single-sideband with suppressed carrier (e.g. Shortwave utility and amateur stations)
K	Pulse amplitude modulation
L	Pulse width modulation (e.g. as used by WWVB)
M	Pulse position modulation
P	Sequence of pulses without modulation
Q	Sequence of pulses, phase or frequency modulation within each pulse
R	Single-sideband with reduced or variable carrier
V	Combination of pulse modulation methods
W	Combination of any of the above
X	None of the above

The second character, almost always a number, tells you the type of modulating signal being applied.

Character	Description
Ø	No modulating signal
1	One channel containing digital information, no subcarrier
2	One channel containing digital information, using a subcarrier

Character	Description
3	One channel containing analog information
7	More than one channel containing digital information
8	More than one channel containing analog information
9	Combination of analog and digital channels
X	None of the above

Types 4 and 5 were removed from use with the 1982 ITU Radio Regulations. In previous editions, they had indicated facsimile and video, respectively.

The third character indicates the type of information to be carried by the modulated signal.

Character	Description
N	No transmitted information
A	Aural telegraphy, intended to be decoded by ear, such as Morse code
B	Electronic telegraphy, intended to be decoded by machine (radioteletype and digital modes)
C	Facsimile (still images)
D	Data transmission, telemetry or telecommand (remote control)
E	Telephony (voice or music intended to be listened to by a human)
F	Video (television signals)
W	Combination of any of the above
X	None of the above

Two more symbols can be used in the ITU system to add more detail, but the FCC does not use them in Part 97, nor do they always use the bandwidth indicator. The fourth character, if present, gives more details of the information to be transmitted.

Character	Description
A	Two-condition code, elements vary in quantity and duration
B	Two-condition code, elements fixed in quantity and duration
C	Two-condition code, elements fixed in quantity and duration, error-correction included
D	Four-condition code, one condition per "signal element"
E	Multi-condition code, one condition per "signal element"
F	Multi-condition code, one character represented by one or more conditions
G	Monophonic broadcast-quality sound
H	Stereophonic or quadraphonic broadcast-quality sound
J	Commercial-quality sound (non-broadcast)

Character	Description
K	Commercial-quality sound—frequency inversion and-or "band-splitting" employed
L	Commercial-quality sound, independent FM signals, such as pilot tones, used to control the demodulated signal
M	Greyscale images or video
N	Full-color images or video
W	Combination of two or more of the above
X	None of the above

The fifth character indicates the type of multiplexing used.

Character	Description
C	Code-division (excluding spread spectrum)
W	Combination of Frequency-division and Time-division
F	Frequency-division
X	None of the above
N	None used
T	Time-division

Common Ham Radio Emission Types

Common Ham Radio Emission Types	
Emission Designator	What it Means
A1A	CW
A2A	Modulated tones on a carrier
A3E	Double-sideband AM
C3F	NTSC analog television
F1C	FSK
J2B	Phase shift keying, such as PSK31
J3E	SSB Phone
R3E	SSB with reduced or variable carrier
F3E	FM Phone

Wire Size and Ampacity Guidelines

WARNING! Installation of electrical wire can be hazardous and, if done improperly can result in personal injury or property damage. For safe wiring practices, consult the National Electrical Code®, your local building inspector, or a qualified electrician.

Ampacities are based on the 2017 NEC and do not reflect any temperature correction or ampacity adjustments that may be required. Please consult a qualified electrician or professional engineer to determine the appropriate values for your specific application.

The allowable values are based on temperature alone and do not take voltage drop into consideration.

	Copper			Aluminum	
Temperature Rating	60°C (140°F)	75°C (167°F)	90°C (194°F)	75°C (167°F)	90°C (194°F)
Wire Type (See below)	NM-B[1], UF-B[2]	THW[3], THWN[3], SE[5], USE[6], XHHW[4]	THWN-2[8,3], THHN9[3], XHHW-2[10,4], USE-2[11,6]	THW[3], THWN[3], SE[5], USE[6], XHHW[4]	XHHW-2[4], THHN[3], THWN-2[3]
Wire Gauge Size					
14	15	20	25	---	---
12	20	25	30	20	25
10	30	35	40	30	35
8	40	50	55	40	45
6	55	65	75	50	55
4	70	85	95	65	75
3	85	100	115	75	85
2	95	115	130	90	100
1		130	145	100	115
1/0		150	170	120	135
2/0		175	195	135	150
3/0		200	225	155	175

	Copper			Aluminum	
Temperature Rating	60°C (140°F)	75°C (167°F)	90°C (194°F)	75°C (167°F)	90°C (194°F)
Wire Type (See below)	NM-B[1], UF-B[2]	THW[3], THWN[3], SE[5], USE[6], XHHW[4]	THWN-28[3], THHN9[3], XHHW-210[4], USE-211[6]	THW[3], THWN[3], SE[5], USE[6], XHHW[4]	XHHW-2[4], THHN[3], THWN-2[3]
Wire Gauge Size					
4/0		230	260	180	205
250		255	290	205	230
300		285	320	230	260
350		310	350	250	280
500		380	430	310	350
600		420	475	340	385
750		475	535	385	435
1000		545	615	445	500

Wire Types

1) NM-B: **N**on-**m**etallic jacket cable, indoor use. Commonly called "Romex."

2) UF-B: **U**nderground **f**eeder. It has added protection to allow it to be buried in the ground without any additional conduit, raceway or tray.

3) THW, THHN, THWN: Single strand, **t**hermoplastic, **h**igh **h**eat, **n**ylon coated. THHN wire without a dual approval of THWN is not water resistant. The added "W" stands for water resistant; it could be installed outdoors and in conduit while THHN wire could not.

4) XHHW: Single strand, **cross-linked** polymer, **h**igh **h**eat, **w**ater-resistant. XHHW-2 is a newer, upgraded version of XHHW.

5) SE: Service Entrance cable.

6) USE and USE-2: **U**nderground **s**ervice **e**ntrance cable. Can be buried. USE-2 is the newer, upgraded version of USE, with a higher temperature rating.

Electrical Wire Run Lengths vs. Voltage Drop

This chart shows maximum wire run lengths to carry a given amperage through a wire of the gauges listed without a more than 2% voltage drop. For instance, if you plan to install a 25-amp circuit using #10 gauge wire, your run must be no longer than 45 feet.

For copper wire, solid, two-conductor, K-11 (77° - 121° F)						
@ 120 V, single phase, maximum 2% voltage drop.						
Maximum Amperage	Volt Amps	Maximum Run Length in Feet #14 AWG	Maximum Run Length in Feet #12 AWG	Maximum Run Length in Feet #10 AWG	Maximum Run Length in Feet #8 AWG	Maximum Run Length in Feet #6 AWG
1	120	450	700	1100	1800	2800
5	600	90	140	225	360	575
10	1200	45	70	115	180	285
15	1800	30	47	75	120	190
20	2400		36	57	90	140
25	3000			45	72	115
30	3600			38	60	95
40	4800				45	72
50	6000					57
@ 240 V, single phase, maximum 2% voltage drop.						
Maximum Amperage	Volt Amps	Maximum Run Length in Feet #14 AWG	Maximum Run Length in Feet #12 AWG	Maximum Run Length in Feet #10 AWG	Maximum Run Length in Feet #8 AWG	Maximum Run Length in Feet #6 AWG
1	240	900	1400	2200	3600	5600
5	1200	180	285	455	720	1020
10	2400	90	140	225	360	525
15	3600	60	95	150	240	350
20	4800		70	110	180	265
25	6000			90	144	210
30	7200			75	120	175
40	5600				90	130
50	12000					105

Values shown are *approximate* and for comparison and estimation purposes only. Consult your local electrical code and/or the National Electrical Code before installing wiring.

Coaxial Cable Specifications – Power Capacity

	Power Capacity (In watts at 104°F, 40°C)							
MHz	30	50	150	220	450	900	1500	2000
RG-174		~197		~135	120	109		
LMR-100A®	230	180	100	80	60	40	30	25
RG-58U	400	300	160		80			
LMR-200®	1020	790	450	370	260	180	140	120
RG-59	500	400	250					
RG-8X	350	280	150		80			
LMR-240®	1490	1150	660	540	380	260	200	170
LMR-240 Ultra®	1490	1150	660	540	380	260	200	170
RG-213	1800	1200	620		300			
RG-214	1800	1200	620		300			
LMR-400®	2100	1700	1000	830	550	380	290	250
LMR-400 Ultra®	2100	1700	1000	830	550	380	290	250
DRF-400	3300	2570	1470	1200	830	580	440	370
Bury-FLEX™	No published data.							
Belden RF300 (Direct bury)	2161	1658	928	758	522	367	278	239
9086		1500	~830	~685	~425	~230		
9913	2200	1700	900		450	280	200	160

Values shown are *approximate* and for comparison purposes only.
LMR® is a registered trademark of Times Microwave Systems.

These ratings have been gathered from various sources, including manufacturers' data sheets. In some cases no data was available from any credible source, so none is reported here. In other cases, values were estimated from published data; such values are indicated by a "~." Cables with similar or even identical names but from different manufacturers often have different characteristics.

No recommendation or endorsement is implied by these listings.

Coaxial Cable Specifications - Attenuation

	Attenuation (dB per 100 feet)								
MHz	30	50	100	146	150	440	450	1000	2400
RG-174	5.5	6.6	8.8	13.0		25.0		30.0	75.0
LMR-100A®	3.9	5.1		8.8	8.9	15.6	15.8		
RG-58A/U	2.5	4.1	5.3	6.1	6.1	10.4	10.6	24.0	38.9
LMR-200®	1.8	2.3		3.9	4.0	6.9	7.0		16.5
RG-59		2.4	3.5			7.6		12.0	
RG-8X	2.0	2.1	3.0	4.5	4.7	8.1	8.6		21.6
LMR-240®	1.3	1.7		3.0	3.0	5.2	5.3		12.7
LMR-240 Ultra®	1.3	1.7		3.0	3.0	5.2	5.3		12.7
RG-8/U FOAM		1.2	1.8					7.1	
RG-213		1.5	2.1	2.8	2.8	5.1	5.1	8.2	
RG-214	1.2	1.6	1.9	2.8	2.8	5.1	5.1	8.0	13.7
LMR-400®	0.7	0.9		1.5	1.5	2.7	2.7		6.6
LMR-400 Ultra®	0.7	0.9		1.5	1.5	2.7	2.7		6.6
DRF-400	0.7	0.9		1.5			2.7		6.7
Bury-FLEX™		1.1	1.5					4.8	
Belden RF300 (Direct bury)	1.0	1.3			2.2		3.9		
9086			1.4			2.8	2.8		
9913	0.8			1.5		2.8			7.5

Values indicated are *approximate* and for comparison purposes only.
LMR® is a registered trademark of Times Microwave Systems.

These ratings have been gathered from various sources, including manufacturers' data sheets. In some cases no data was available from any credible source, so none is reported here. In other cases, values were estimated from published data; such values are indicated by a "~." Cables with similar – even identical – names but from different manufacturers often have different characteristics.

No recommendation or endorsement is implied by these listings.

Schematic Symbols

Note: Symbols shown are IEEE standard. Not all users follow these standards, and even the IEEE specifications contain variants of these symbols.

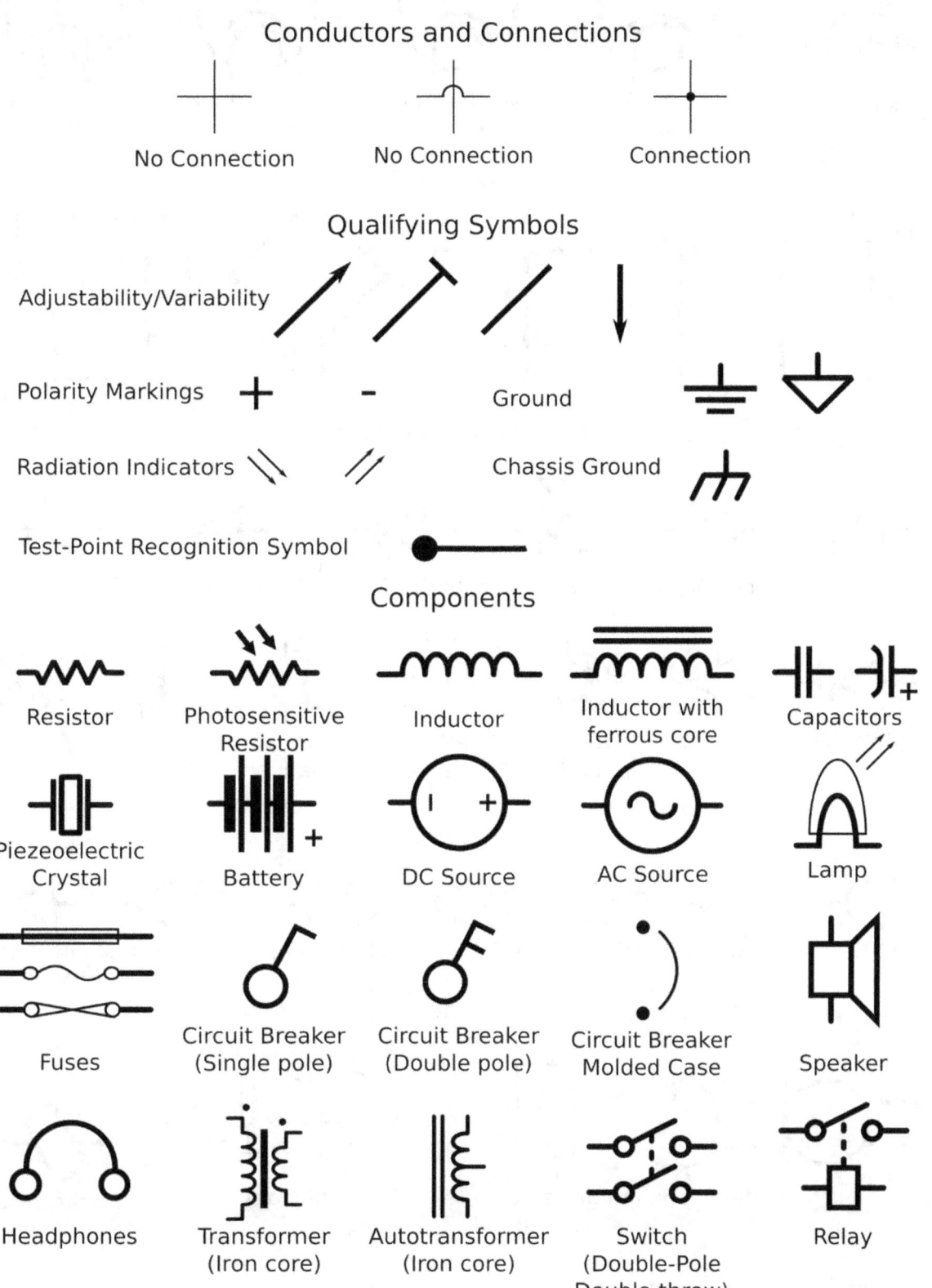

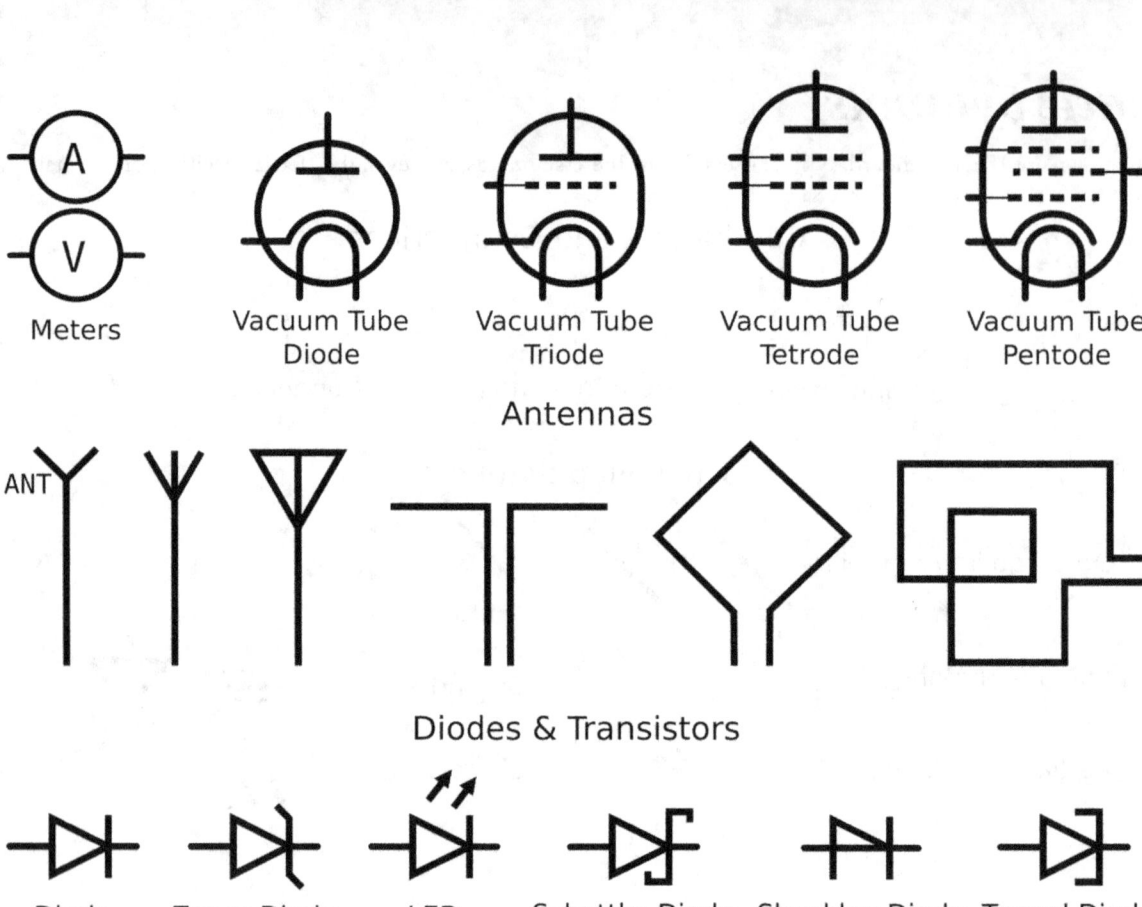

Antennas

Diodes & Transistors

Logic Gates

Other

Antenna Lengths

All values given are approximations for resonant dipole antennas made of copper wire, mounted ¼ to ½ wavelength above ground level. Different wire gauges, different heights, and even different wire materials may require different lengths to achieve resonance.

It is easy to trim a wire shorter and challenging to trim it longer. Be wise.

These figures do not include surplus for connections to insulators, etc.

Band	Center Frequency MHz	1/4 Wavelength Feet	1/2 Wavelength Feet	1/2 Wavelength Inverted V, 90 Feet
2200	0.137	1708.03	3416.06	3381.90
630	0.475	492.63	985.26	975.41
160	1.800	130.00	260.00	257.40
	1.850	126.49	252.97	250.44
	1.900	123.16	246.32	243.85
	2.000	117.00	234.00	231.66
80	3.550	65.92	131.83	130.51
	3.750	62.40	124.80	123.55
	3.850	60.78	121.56	120.34
	3.900	60.00	120.00	118.80
60	5.370	43.58	87.15	86.28
40	7.070	33.10	66.20	65.53
	7.212	32.45	64.89	64.24
	7.238	32.33	64.66	64.01
30	10.100	23.17	46.34	45.87
	10.125	23.11	46.22	45.76
	10.150	23.05	46.11	45.65
17	18.089	12.94	25.87	25.61
	18.140	12.90	25.80	25.54
15	21.100	11.09	22.18	21.96
	21.325	10.97	21.95	21.73
12	24.910	9.39	18.79	18.60
	24.960	9.38	18.75	18.56
10	28.150	8.31	16.63	16.46
	29.000	8.07	16.14	15.98
6	50.050	4.68	9.35	9.26
	52.000	4.50	9.00	8.91
2	144.000	1.63	3.25	3.22
	146.000	1.60	3.21	3.17
	148.000	1.58	3.16	3.13

Optimum "Random" Wire Antenna Lengths

Random wire antennas are end-fed wire antennas of random length. There are, however, particular lengths of wire that will optimize performance of your random wire antenna. Wire lengths near an even multiple of ½ wavelengths are problematic, while those whose lengths are odd multiples of ½ wavelengths are more effective.

These values are given as examples and for estimation purposes. Calculate the length for the center of your desired band by using the formula below. (Some values are given strictly for entertainment purposes. It's unlikely anyone is going to build a 1, 813 ft long antenna for 160 meters, nor a 1.13 ft long one for 70 cm.

Much credit to Patrick Lambert, WØIPL, for all his work on random wire antennas.

Random Wire for Center of Band – Use Odd Multiples of ½ Wavelengths for Best Results (Avoid lengths that are even multiples of ½ wavelengths)							
	160 Meters	80 Meters	75 Meters	40 Meters	60 Meters	30 Meters	20 Meters
Center Frequency in MHz	1.900	3.550	3.800	7.175	5.367	10.125	14.175
Number of 1/2 wavelengths	ft	ft	ft	ft	ft	ft	ft
1	259.01	138.63	129.51	68.59	91.69	48.60	34.72
2	518.02	277.25	259.01	137.18	183.39	97.21	69.43
3	777.03	415.88	388.52	205.76	275.08	145.81	104.15
4	1036.04	554.50	518.02	274.35	366.77	194.42	138.87
5	1295.05	693.13	647.53	342.94	458.47	243.02	173.59
6	1554.06	831.75	777.03	411.53	550.16	291.63	208.30
7	1813.07	970.38	906.54	480.12	641.86	340.23	243.02
	17 Meters	15 Meters	12 Meters	10 Meters	6 Meters	2 Meters	70 cm
Center Frequency in MHz	18.118	21.225	24.940	28.850	52.000	146.000	435.000
Number of 1/2 wavelengths	ft	ft	ft	ft	ft	ft	ft
1	27.16	23.19	19.73	17.06	9.46	3.37	1.13
2	54.32	46.37	39.46	34.12	18.93	6.74	2.26
3	81.49	69.56	59.20	51.17	28.39	10.11	3.39
4	108.65	92.74	78.93	68.23	37.86	13.48	4.53
5	135.81	115.93	98.66	85.29	47.32	16.85	5.66
6	162.97	139.12	118.39	102.35	56.78	20.22	6.79
7	190.13	162.30	138.13	119.41	66.25	23.59	7.92

To calculate for your desired center frequency: $\frac{150}{frequency} \times 3.28 \times number\ of\ \frac{1}{2}\ wavelengths\ desired$.

For an "all band" (75 – 10) random wire, Patrick recommends a 74 foot wire. (You'll also need an antenna tuner and a solid connection to ground for the other side of the transmitter output.)

Power Loss from SWR

SWR	% Power Loss	dB Loss	ERP
1.1:1	0.23%	0.01 dB	99.77%
1.2	0.83%	0.04 dB	99.17%
1.3	1.70%	0.07 dB	98.30%
1.4	2.78%	0.12 dB	97.22%
1.5	4.00%	0.18 dB	96.00%
1.6	5.33%	0.24 dB	94.67%
1.7	6.72%	0.30 dB	93.28%
1.8	8.16%	0.37 dB	91.84%
1.9	9.63%	0.44 dB	90.37%
2	11.11%	0.51 dB	88.89%
2.1	12.59%	0.58 dB	87.41%
2.2	14.06%	0.66 dB	85.94%
2.3	15.52%	0.73 dB	84.48%
2.4	16.96%	0.81 dB	83.04%
2.5	18.37%	0.88 dB	81.63%
2.6	19.75%	0.96 dB	80.25%
2.7	21.11%	1.03 dB	78.89%
2.8	22.44%	1.10 dB	77.56%
2.9	23.73%	1.18 dB	76.27%
3	25.00%	1.25 dB	75.00%
3.5	30.86%	1.60 dB	69.14%
4	36.00%	1.94 dB	64.00%
4.5	40.50%	2.25 dB	59.50%
5	44.44%	2.55 dB	55.56%
5.5	47.93%	2.83 dB	52.07%
6	51.02%	3.10 dB	48.98%
6.5	53.78%	3.35 dB	46.22%
7	56.25%	3.59 dB	43.75%
7.5	58.48%	3.82 dB	41.52%
8	60.49%	4.03 dB	39.51%
8.5	62.33%	4.24 dB	37.67%
9	64.00%	4.44 dB	36.00%
9.5	65.53%	4.6 dB	34.47%
10	66.94%	4.8 dB	33.06%
11	69.44%	5.1 dB	30.56%
12	71.60%	5.5 dB	28.40%
13	73.47%	5.8 dB	26.53%
14	75.11%	6.0 dB	24.89%
15	76.56%	6.3 dB	23.44%

RF Exposure Limits

When your station operates above certain power levels at certain frequencies, you must do a formal RF exposure evaluation of your station.

RF Exposure Evaluation Limits	
Band	Evaluation Required if Power Exceeds:
80, 75 & 40 meters	500 watts
30 meters	425 watts
20 meters	225 watts
17 meters	125 watts
15 meters	100 watts
10 meters	50 watts
6, 2, and 1.25 meters	50 watts
70 cm	70 watts
33 cm	150 watts
23 cm	200 watts
13 cm and higher	250 watts
Repeater stations on all bands	Non-building mounted antennas: Height above ground level to lowest point of antenna less than 10 meters (32.8 feet) *and* power greater than 500 watts *ERP*. Building-mounted antennas: Power greater than 500 watts *ERP*.

Worksheets and information to manually complete an RF exposure evaluation: http://www.arrl.org/rf-exposure

World Voltages

Country	Voltage	Frequency	Country	Voltage	Frequency	Country	Voltage	Frequency
Australia	230V	50Hz	Ireland	230V	50Hz	South Africa	220V	50Hz
Brazil	117V or 220V	60Hz	Israel	230V	50Hz	Thailand	220V	50Hz
Canada	120V	60Hz	Italy	230V	50Hz	UK	230V	50Hz
China	220V	50Hz	Japan	100V	50/60Hz	USA	120V	60Hz
France	230V	50Hz	New Zealand	230V	50Hz			
Germany	230V	50Hz	Philippines	220V	60Hz			
India	230V	50Hz	Russia	220V	50Hz			

Electrical Values & Abbreviations

Symbol	Units	Value
Ω	ohms	Resistance, Reactance, and Impedance
ϱ	ohms per meter	Resistivity
Φ_E*	volt meters (V m)	Electric flux
μ	henrys per meter (H/m)	Permeability
λ*	meters	Wavelength
A	amperes	Electric Current
A/m²	amperes per square meter	Electric Current Density
B	siemens (formerly mhos)	Inverse of impedance
C	coulomb	Electric Charge
C/m²	coulombs per square meter.	Electric displacement field
E* Also: ΔV*, Δφ*, U*	volts	Potential difference; electromotive force
F	farads	Capacitance
f*	hertz	Frequency
G	siemens (formerly mhos)	Conductance
H	henrys	Inductance
Hz	hertz	Frequency (cycles per second)
I*	amperes	Electric current
J	amperes per square meter (A/m²)	Electric current density
j*	none	The imaginary number: $\sqrt{-1}$
L*, sometimes M*	henrys	inductance
Q*	coulombs	Electric Charge
P*	watts	Power
R*	ohms	Resistance
S	siemens	Conductivity; Admittance &; Acceptance
T	teslas	Magnetic flux density; Magnetic induction
W	watts	Power
Wb	webers	Magnetic flux
X*	ohms	Reactance
Y*	siemens	Admittance
Z*	ohms	Impedance

*Used as values in formulas. i.e.: E = I x R.

Metric Electronic Values

Divide by Power of 10 ←

Prefix	Symbol	Example	Multiplier	Scientific Notation
Standard Values			1	10^0
Giga	G	GHz	1,000,000,000	10^9
Mega	M	MHz	1,000,000	10^6
kilo	k	kV	1,000	10^3
deci	d	dB	0.1	10^{-1}
milli	m	mA	0.001	10^{-3}
micro	μ	μF	0.000001	10^{-6}
nano	n	nF	0.000000001	10^{-9}
pico	p	pF	0.000000000001	10^{-12}

→ **Multiply by Power of 10**

To convert prefixed values to standard values, **multiply** the prefixed value by the scientific notation power of 10 value for that prefixed value. To convert milliamperes to amperes, multiply the number of milliamperes by 10^{-3}.

To convert standard values to prefixed values, **divide** the standard value by the scientific notation power of 10 value. To convert amperes to milliamperes, divide the number of amperes by 10^{-3}.

Ah, but how to remember those prefixed values? It's not as daunting as you might think. First, notice, that except for "deci", all of those exponents can be divided by three. -12, -9, -6, -3, 3, 6, and 9 all divide evenly by three. That's not an accident. There's a slight difference between scientific notation and **engineering notation** and that's simply that all exponents in engineering notation are divisible by 3. So in scientific notation, we'd write 30, 000 as 3×10^4, but in engineering notation it would be 30×10^3. "Deci" is the oddball, at 10^{-1}, but that's fine, there's just one odd one to keep track of and in almost all cases we use that one for only one value, the "decibel."

To keep track of those prefix values, memorize this shocking phrase about insane jealousy in the ham community: "**P**AUL'S **12** **N**EW **MICRO**PHONES **M**ADE **DESI** **K**ILL **O**LD **M**AD **G**EORGE." The "P" of Paul stands for pico. The 12 reminds you that pico is 10^{-12}. "New" stands for nano, "**micro**phone" for micro, "made" for milli, "Desi" for deci, "kill old" for kilo, "Mad" for mega, and "George" for giga. The exponent values start at -12 and move up by threes through -9, -6, and -3 until we get to the odd one, deci at -1, then 3, 6, and 9 for kilo, mega, and giga.

Note: In old schematics, textbooks, and articles, "mm", for "micro-micro" is often substituted for "pico." 1 mmfd = 1 picofarad.

Resistor Color Codes

Band Color		Value			Multiplier		Tolerance		Temperature Coefficient	
Color	Code	Band 1	Band 2	Band 3			%	Letter	ppm/K	Letter
Black	BK	0	0	0	$\times 10^0$	1			250	U
Brown	BN	1	1	1	$\times 10^1$	10	± 1%	F	100	S
Red	RD	2	2	2	$\times 10^2$	100	± 2%	G	50	R
Orange	OG	3	3	3	$\times 10^3$	1,000			15	P
Yellow	YE	4	4	4	$\times 10^4$	10,000			25	Q
Green	GN	5	5	5	$\times 10^5$	100,000	± 0.5%	D	20	Z
Blue	BU	6	6	6	$\times 10^6$	1,000,000	± 0.25%	C	10	Z
Violet	VT	7	7	7	$\times 10^7$	10,000,000	± 0.1%	B	5	M
Gray	GY	8	8	8	$\times 10^8$	100,000,000	± 0.05%	A	1	K
White	WH	9	9	9	$\times 10^9$	1,000,000,000				
Gold	GD				$\times 10^{-1}$	0.1	± 5%	J		
Silver	SR				$\times 10^{-2}$	0.01	± 10%	K		
Pink	PK				$\times 10^{-3}$	0.001				
None							± 20%	M		

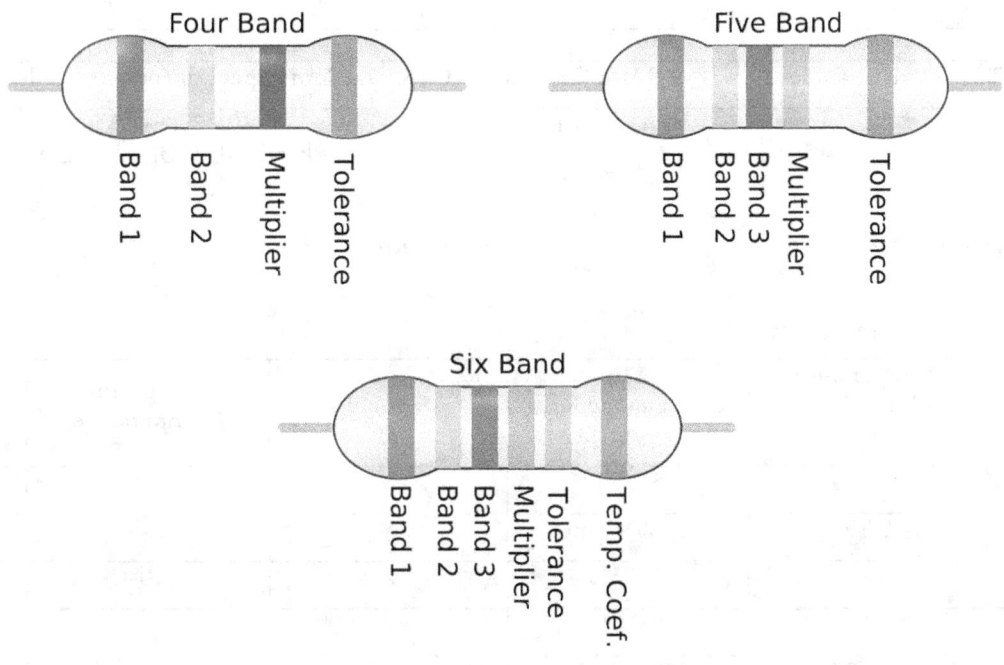

Mica Capacitor Color Code

Codes for 3, 5 & 6 Dot Capacitors (Check back of capacitor for extra codes)

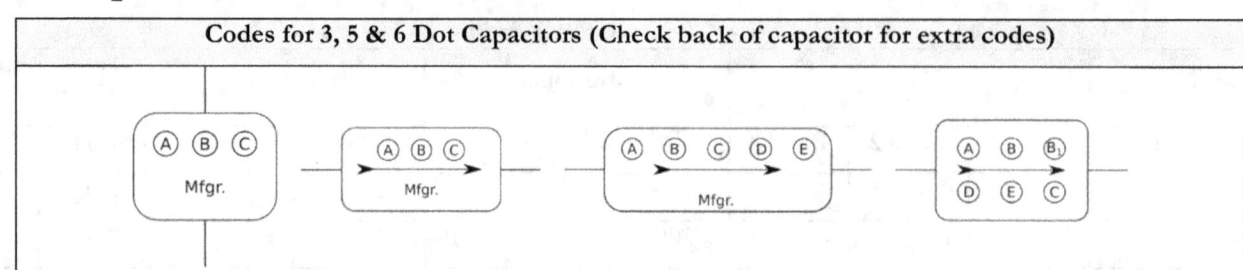

Color	Dots A, B, B₁ 1st, 2nd, 3rd figure	Dot C # of 0's after 2nd figure	Dot D Voltage Rating	Dot E Tolerance
Black	0	0		Black or No Color = ± 20%
Brown	1	1	Brown = 100	Brown = ± 1%
Red	2	2	Red = 200	Red = ± 2%
Orange	3	3	Orange = 300	Orange = ± 3%
Yellow	4	4	Yellow = 400	Yellow = ± 4%
Green	5	5	Green = 500	Green = ± 5%
Blue	6	6	Blue = 600	Blue = ± 6%
Violet	7	7	Violet = 700	Violet = ± 7%
Gray	8	8	Gray = 800	Gray = ± 8%
White	9	9	White = 900	White = ± 9%
Silver				Silver = ± 10%
Gold			1000 (EIA)	Gold = ± 0.5 %

Characteristic Codes for 7, 8 & 9 Dot Capacitors (Check back of capacitor for extra codes.

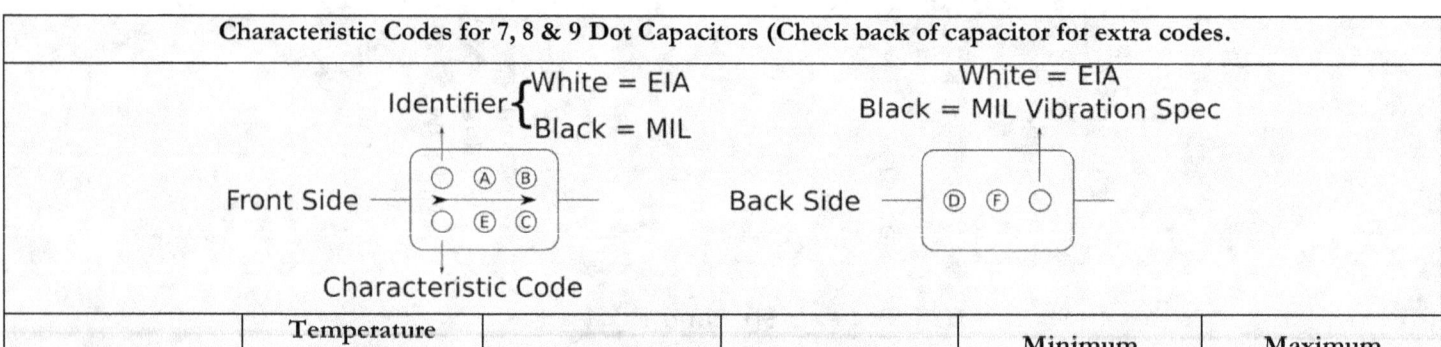

Color	Temperature Coefficient (ppm/°C)	Capacitance Drift	Symbol	Minimum Temperature	Maximum Temperature
Black	± 1000	± 5% +1pF	X	-55°C	
Brown	± 500	± 3% + 1 pF	Y	-30°C	
Red	± 200	± 0.5%	Z	+10°C	
Orange	± 100	± 0.3%	2		+45°C
Yellow	-20 to + 100	± 0.1% + 0.1 pF	4		+65°C
Green	0 to +70	± 0.05% + 0.1 pF	5		+85°C
			6		+105°C
			7		+125°C

Symbol	Max. Cap. Change over Temp. Range	Symbol	Max. Cap. Change over Temp. Range	Symbol	Max. Cap. Change over Temp. Range
A	± 1.0%	E	± 4.7%	S	± 22%
B	± 1.5%	F	± 7.5%	T	-33%, +22%
C	± 2.2%	P	± 10%	U	-56%, +22%
D	± 3.3%	R	± 15%	V	-82%, +22%

Ohm's Law

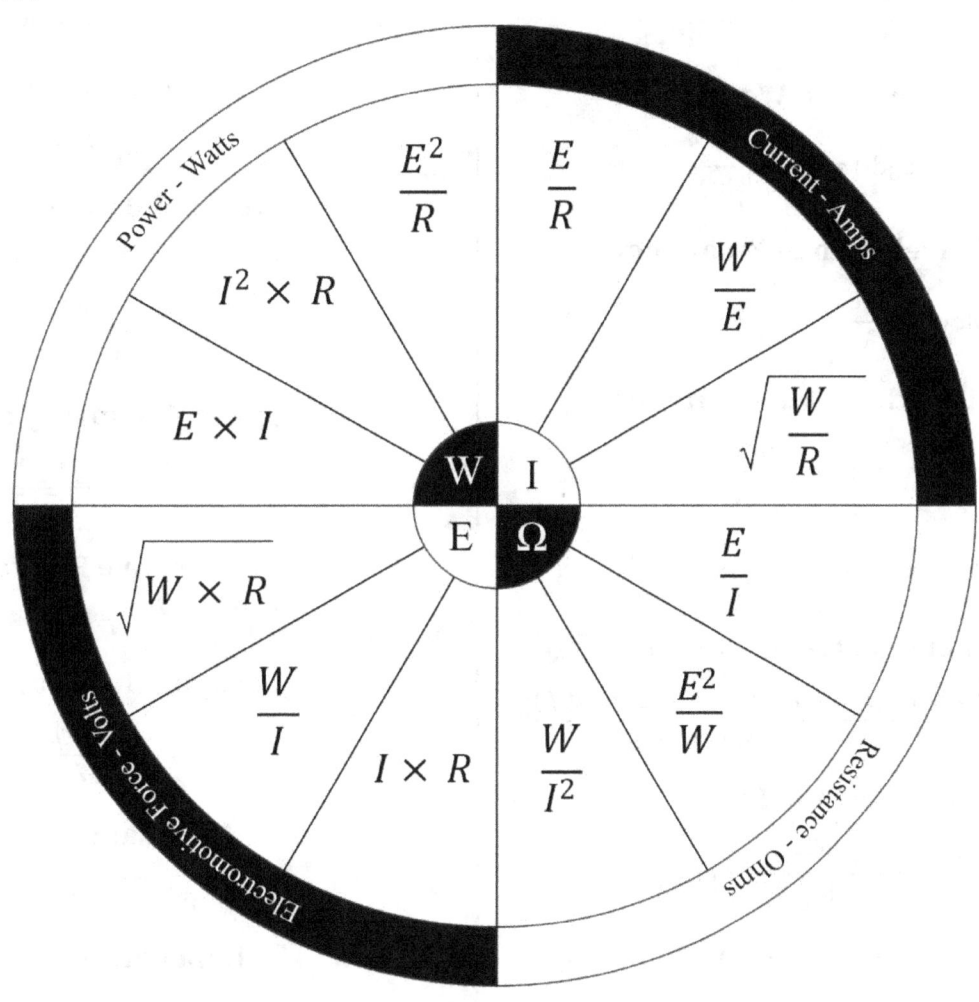

Electronic Formulas

Series Resistance, Series Inductance, and Parallel Capacitance

$$Total = Value_1 + Value_2 \ldots Value_n$$

Parallel Resistance, Parallel Inductance, and Series Capacitance

$$Total = \frac{1}{\frac{1}{Value_1} + \frac{1}{Value_2} \ldots \frac{1}{Value_n}}$$

Convert Frequency to Wavelength

$$\lambda\ (Wavelength) = \frac{300}{Frequency}$$

Convert Wavelength to Frequency

$$Frequency = \frac{300}{\lambda}$$

Antenna Length, ½ Wave Antenna

$$Length\ (feet) = \frac{486}{f_{MHz}}$$

$$Length\ (inches) = \frac{5832}{f_{MHz}}$$

Adjusting Length of Dipole Antenna

$$New\ Length = \frac{(Present\ Length \times Present\ Resonant\ f)}{Desired\ Resonant\ f}$$

Power Loss from SWR

$$Watts\ Lost = \frac{(SWR-1)^2}{(SWR+1)^2} \times PEP$$

Antenna Length, ¼ Wave Antenna

$$Length\ (feet) = \frac{243}{f_{MHz}}$$

$$Length\ (inches) = \frac{2916}{f_{MHz}}$$

Note: These formulas are approximations for dipoles at ¼ to ½ wavelength above ground and assume you will trim the antenna to the proper length.

Percent Change of Power Increase from Decibels

$$\%\ Increase = 100 \times 10^{\left(\frac{dB}{10}\right)}$$

Decibels from Power Change

$$dB = 10\ log\left(\frac{Power_1}{Power_2}\right)$$

Percent Power Remaining After dB Change

$$\%\ Power\ Remaining = 100 - \left(100 \times \frac{1}{10^{\left(\frac{dB}{10}\right)}}\right)$$

dB to Power Ratio

$$Power\ Ratio = 10^{(dB \div 10)}$$

Power Ratio to dB

$$dB = 10_{log}(Power\ Ratio)$$

Intermodulation Frequency

$$f_{imd} = 2f_1 - f_2$$

and

$$f_{imd} = 2f_2 - f_1$$

Resonant Frequency of a Circuit

$$f = \frac{1}{2\pi\sqrt{L \times C}}$$

Inductive Reactance

$$X_L = 2\pi f L$$

Capacitive Reactance

$$X_C = \frac{1}{2\pi f C}$$

Impedance (Series)

$$Z = \sqrt{R^2 + (X_L - X_C)^2}$$

Impedance (Parallel)

$$Z = \frac{R \times (X_L - X_C)}{\sqrt{R + (X_L - X_C)}}$$

Q of a Series Resonant Circuit

$$Q = \frac{X}{R}$$

Q of a Parallel Resonant Circuit

For a parallel circuit with R in series or for a small value of R in parallel:

$$Q = \frac{X}{R}$$

For a parallel circuit with a large value of R in parallel:

$$Q = \frac{R}{X}$$

Half power bandwidth

$$\text{half power bandwidth} = \frac{f_{resonant}}{Q}$$

Time Constant of a Capacitor

$$\tau = RC$$

Time Constant of an Inductor

$$\tau = \frac{L}{R}$$

Susceptance

$$\text{Susceptance} = \frac{X_L - X_C}{R^2 + (X_L - X_C)^2}$$

Or more simply ....

$$B = \frac{1}{Z}$$

Phase Angle

$$\text{Phase Angle} = \tan^{-1}\left(\frac{X_L - X_c}{R}\right)$$

Power Factor

$$\text{Power Factor} = \cos(\text{Phase Angle})$$

RMS Voltage from Peak Voltage

$$V_{RMS} = V_{Peak} \times 0.707$$

or, more precisely

$$V_{RMS} = \frac{V_{Peak}}{\sqrt{2}}$$

Peak Voltage from RMS Voltage

$$V_{RMS} = \frac{V_{Peak}}{0.707}$$

Peak Voltage from Peak-to-Peak

$$V_{Peak} = \frac{V_{Peak-to-peak}}{2}$$

Turns on a Coil for a Desired Inductance

$$N = \sqrt{\frac{L}{A_L}}$$

Where A_L is the inductance index of the core material. While this is the Extra Class license exam's formula, in practice it is, at best, an approximation. For an air core single-layer inductor, a more precise formula is:

$$N = \sqrt{\frac{L \times \ell}{\mu_0 \times A}}$$

Where L is the desired inductance, ℓ is the length of the wire in meters, μ_0 is the permeability of free space ($4 \times \pi \times 10^{-7}$), and A is the inner core area in square meters.

For more complex coils, the reader is advised to consult any of the numerous on line coil calculators.

Voltage at the Secondary Winding of a Transformer Based on Turns (N)

$$Vs = \frac{N_S}{N_P} \times Vp$$

Power Dissipation of a Series Connected Linear Voltage Regulator

$$(E_{input} - E_{output}) \times I_{output}$$

Gain of an Op-amp

$$\text{Gain} = \frac{R_F}{R_1}$$

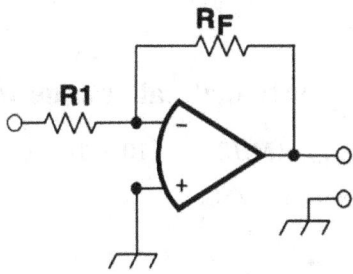

Modulation Index (Deviation ratio)

$$\text{Modulation Index} = \frac{\Delta f}{f_{max}}$$

Bandwidth from WPM or Baud rate

For CW: $\text{bandwidth} = WPM \times 4$

For frequency shifted carriers: $\text{bandwidth} = (K \times \text{shift}) + \text{Baud rate}$

For AM: $\text{bandwidth} = \text{Baud rate} \times K$

K is a constant reflecting the acceptable level of distortion and the quality of the signal path. For amateur radio purposes, use 1.2 for K, except for CW when the value of K is 4.8.

Antenna Efficiency

$$Efficiency = (R_{Radiation}/R_{Total}) \times 100\%$$

Impedance Matching Transformer

$$Z_{Transformer} = \sqrt{Z_1 \times Z_2}$$

Bandwidth of an FM Signal
(Carrying multiple frequencies.)

$$Bandwidth = 2(\Delta f + f_m)$$

Δf = Maximum deviation of carrier.

f_m = Maximum frequency to be transmitted.

Calculate Radius of Fresnel Zones

$$Radius_1 = \frac{1}{2}\sqrt{\lambda d} \times 0.6$$

$$Radius_2 = \frac{1}{2}\sqrt{2\lambda d} \times 0.6$$

$$Radius_3 = \frac{1}{2}\sqrt{3\lambda d} \times 0.6$$

λ = frequency, d = distance from transmitter. Impingements on the signal path inside odd number radii produce received minima, impingements inside even number radii produce received maxima.

Calculate Critical Frequency

$$f_{critical}(MHZ) = 9.10 \times 10^{-6}\sqrt{N}$$

N = Free electron density/M^3.

Space Weather Terms

Sunspots and Sunspot Numbers

Sunspot activity is expressed by R, the sunspot number, also known as the Wolf number, but most commonly known in the ham radio world as "the SSN." The number doesn't tell you the number of sunspots, it tells you the relative amount of sunspot activity. The formula is:

$$SSN = k(10g + s)$$

"s" is the number of individual spots seen by the observer, "g" is the number of sunspot groups, and "k" is a constant that varies with the observatory's location and equipment.

The Solar Flux Index

Every day, the Penticton Radio Telescope Observatory, located about 200 miles east of Vancouver, BC, reports the Solar Flux Index. Along with the sunspot number, the SFI is one of our most valuable tools for predicting propagation. The scale, in principle, goes from 62.5 to infinity, but in reality the range is more like 62.5 to 300.

Technically, the SFI is a measure of solar radiation at 10.7 centimeters wavelength. It's reported in Solar Flux Units. Typical solar flux numbers will range from 50 to 300 SFU's. Higher fluxes usually mean a higher MUF, lower numbers a lower MUF, and that overall conditions will not be wonderful. However, it takes some time, like a few days, for a change in solar flux to change conditions in the ionosphere, so Monday's report of a 250 SFI might not make for great DX'ing until Wednesday or Thursday.

There's one SFU value for the whole planet, unlike some other values listed below.

K-index

These indices tell us about the space weather up in the magnetosphere. Is it a calm day up there, or is there a storm going on?

The K-index is measured at various magnetic observatories, so different places report different values. The K-index is, basically, a three-hour snapshot of the state of the magnetic field, so the K-index is a measure of the short term stability of the Earth's magnetic field.

K-index values are *quasi-logarithmic*. They'd be regular old logarithmic except the fact that they're measured at multiple locations around the world at different times of day throws off the math a little. They're (almost) like decibel values, where 3 dB is a doubling of power, 10 dB is 10 times power, etc. K-index values from 0 to 1 indicate a quiet day in the magnetic field. Values of 1 to 5 indicate a minor or moderate solar storm. Values over 5 indicate major storms that severely hamper or black out HF altogether. A 9 would be a genuinely scary event that would wipe out all sorts of electronic devices, not to mention playing havoc with the electrical transmission system.

All the daily K-index values are then manipulated mathematically to create the daily A-index.

A-index

The K-index is the short term stability of the geomagnetic field. The A-index is a measure of the long term stability of the Earth's geomagnetic field. In this case "long term" means "one day." Because of the complex nature of the mathematics applied to create the index, the A-index runs from 0 to 400.

Once the K-index and A-index values are calculated for all the various magnetic observatories, a pair of global indicators are generated, known as Kp and Ap, for "planetary" K and A.

High K (and hence, A) values work *against* high solar flux levels when it comes to radio. We like high solar fluxes with low K's and A's. High K's and A's mean the Maximum Usable Frequency's coming down, maybe even below the Lowest Usable Frequency.

Electron Flux

The solar wind consists mostly of a form of matter called plasma. Plasma is made up of negatively charged free electrons and positively charged ions – and if that sounds familiar, it's because the ionosphere various percentages of plasma. One value NOAA reports daily is the Electron Flux index. This is the density of free electrons in the solar wind. The scale runs from zero to "we don't actually know how high this can go." 1000 is a *big* number, though, and is associated with near total HF blackouts through the polar latitudes. Under 100 has little to no impact on HF. If you're planning some trans-polar DX'ing, this is a good number to keep an eye on.

Proton Flux

The density of protons in the solar wind. Like Electron Flux, the scale goes from 0 to ???. Both proton flux and electron flux most influence the E layer.

Solar Wind

The SW index tells you the speed of the solar wind in kilometers per second. The scale goes up to 1000, but that's extreme. Values over 200 km/sec have some impact on HF communication, with values over 600 having fairly severe impact.

X-ray Flux

The GOES (Geostationary Operating Environment Satellites) satellites hover in geostationary orbit over the Earth, mostly keeping track of weather events down here on Earth. However, they also keep an eye on space weather. Among the things they track is the X-ray Flux index. High X-ray fluxes indicate solar flares, solar coronal holes or even coronal mass ejections – big explosions in the sun's corona which can launch massive amounts of particles. As you might imagine, if those particles hit the magnetosphere, it can affect the ionosphere; usually in a negative way from our point of view.

X-ray flux values have an odd scale that matches up with solar flare values. They run from A 0.0 through A 9.9, then B 0.0 through 9.9, C 0.0 through 9.9, then skip to M 0.0 through 9.9, and X 0.0 through, supposedly, 9.9. B's are 10 times more powerful than A's, and each letter represents another 10x increase. That puts the X level at 10, 000 times the A level. At the X level, each increase of 1.0 is a doubling of the original. Any X is a *big* solar flare – an X 9 is about 5, 000, 000 times an A.

I said "supposedly" about the X 9.9 value because in 2003 a solar flare was recorded that blew past the satellite's ability to measure it at X 28. They think it might have hit X 45.

304A

The 304A solar parameter is the measure of ultraviolet emissions at 304 angstroms, correlated to the solar flux index.

The solar flux index was standardized back in the 1950's – it's a measure of the sun's radiation at the frequency of 2800 MHz, a wavelength of 10.7 cm. Think of it as a broad view of the Sun's activity. Very useful, and now we have daily historical data going back some 60 years for comparison purposes, so the solar flux index isn't going away. However, it was created before we had satellites. We don't get much of the Sun's UV radiation down here on the surface, and the amount varies wildly with atmospheric conditions. But stick a satellite up above the atmosphere, and we can get a super-accurate reading of the activity in the UV parts of the spectrum, and that turns out to be a more accurate and immediately useful measure than the solar flux index.

The "A" in 304 A stands for angstrom units, equal to one 10,000,000th of a millimeter. 304 angstroms is a wavelength of ultraviolet light.

304 A is responsible for most of the ionization of the F layer.

B_z

B_z indicates the direction and strength of the interplanetary magnetic field – the magnetic field between us and the Sun. The interplanetary magnetic field is mostly the Sun's magnetic field. It gets measured and quantified in several ways. Among those ways are the B indices – B_X, B_Y, and B_Z. B_X measures the orientation and strength of the field on the X axis of an imaginary three-dimensional grid, with the X axis defined by the plane of the ecliptic – the plane traced by the Earth's orbit. In plain terms, "left or right as we stand at the North Pole and look at the Sun." B_Y tells us about the Y axis – toward the Sun or away from it -- and B_Z – the one with the most effect on our own magnetosphere's events – shows the direction and strength of the interplanetary magnetic field on the Z axis. In other words, B_Z tells us the direction of the North/South axis of the Sun's magnetic field relative to our own field and how strong it is.

Checking the Space Weather

Space weather information is easily available from multiple internet sources. By all means, check out physicist "Dr. T" Tamitha Skov's Solar Storm Forecast on YouTube, for one.

For all the info put together in "ham friendly" style, as well as lots and lots of information about space weather and propagation in general, you can visit this site put together by Paul Herrman, N0BNH.

http://www.hamqsl.com.

FCC Certified VEC's

Anchorage ARC VEC P.O. Box 1283 Kodiak, AK 99615 P: 907-987-6716 E: vec@kl7aa.org	Jefferson Amateur Radio Club P.O. Box 73665 Metairie, LA 70033 E: w5gad.vec@w5gad.org P: 504-636-8809	Sunnyvale VEC Amateur Radio Club, Inc. P.O. Box 70014 Sunnyvale, CA 94086-0014 P: (408) 255-9000 (24 hours) E: vec@amateur-radio.org
American Radio Relay League (ARRL) 225 Main Street Newington, CT 06111-1494 P: (860) 594-0300 F: (860) 594-0339 E: vec@arrl.org	Laurel Amateur Radio Club, Inc. P. O. Box 146 Laurel, MD 20725-0146 P: (301) 937-0394 (1800-2100 hrs) P: (301) 572-5124 (1800-2100 hrs) E: aa3of@arrl.net	W4VEC P.O. Box 482 China Grove, NC 28023-0482 E: raef@lexcominc.net
Central America CAVEC, Inc. 2751 Christian Ln NE Huntsville, AL 35811 P: 256-288-0392 E: cavec.org@gmail.com	MRAC VEC, Inc. c/o Tom Fuszard 2505 S. Calhoun Rd., #203 New Berlin, WI 53151 E: mracvec@gmail.com	W5YI-VEC P.O. Box 200065 Arlington, TX 76006-0065 P: (817) 860-3800 / (800) 669-9594 E: W5YI-VEC@w5yi.org
Golden Empire Amateur Radio Society (GEARS) P.O. Box 508 Chico, CA 95927-0508 P: (530) 893-9211 E: myw6js@gmail.com	MO-KAN VEC Coordinator 228 Tennessee Road Richmond, KS 66080-9174 P: (785) 867-2011 E: Bill.wo0e@gmail.com	Western Carolina Amateur Radio Society VEC, Inc. 7 Skylyn Ct Asheville, NC 28806-3922 P: (828)-253-1192 E: wcarsvec@wcarsvec.net
Greater L.A. Amateur Radio Group(GLAARG) PO Box 500133 Palmdale, CA 9351 661-264-1863 E: vec@glaarg.org	Sandarc-VEC 5511 Marlyland Ave La Mesa, CA 91942-1519 P: (619) 465-3926 E: n6nyx@arrl.net	

Smith Chart

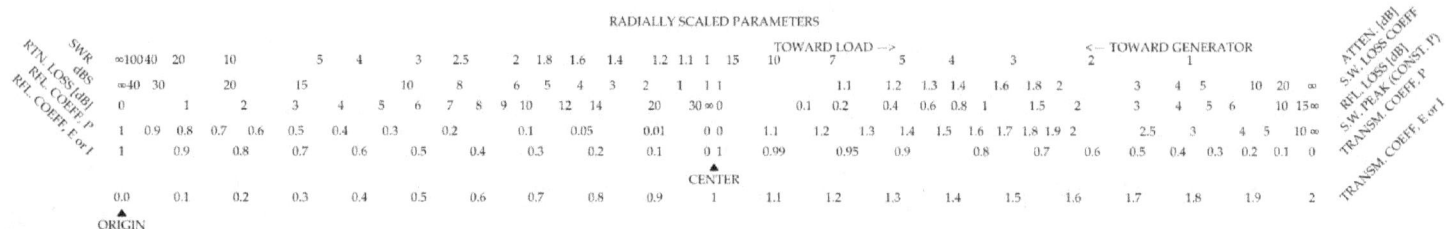

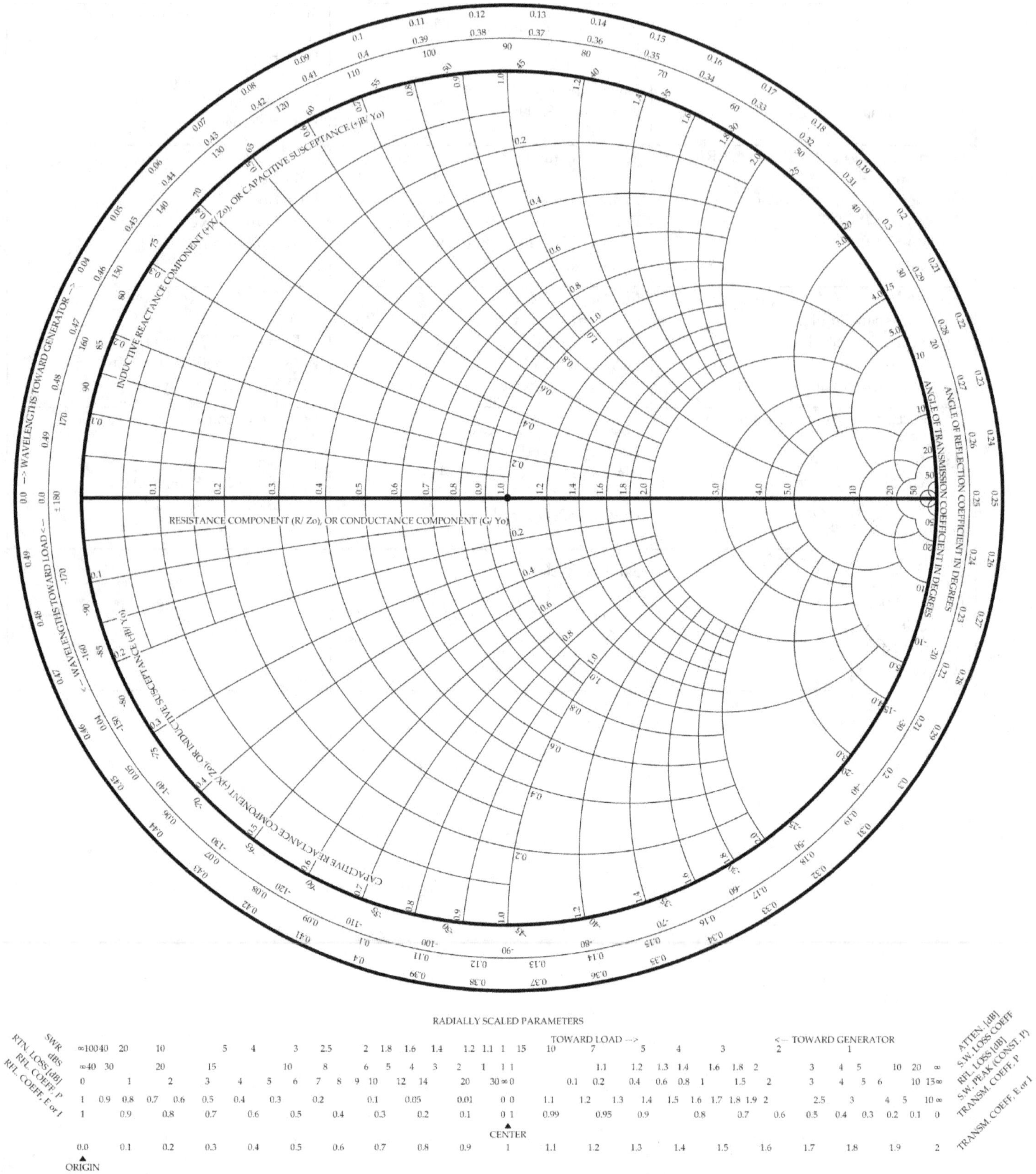

118

FCC Rules & Regulations
Part 97—Amateur Radio Service

AUTHORITY: 47 U.S.C. 151 - 155, 301 - 609, unless otherwise noted.

SOURCE: 54 FR 25857, June 20, 1989, unless otherwise noted.

EDITORIAL NOTE: Nomenclature changes to part 97 appear at 63 FR 54077, Oct. 8, 1998.

Subpart A—General Provisions

§97.1 Basis and purpose.

The rules and regulations in this part are designed to provide an amateur radio service having a fundamental purpose as expressed in the following principles:

(a) Recognition and enhancement of the value of the amateur service to the public as a voluntary noncommercial communication service, particularly with respect to providing emergency communications.

(b) Continuation and extension of the amateur's proven ability to contribute to the advancement of the radio art.

(c) Encouragement and improvement of the amateur service through rules which provide for advancing skills in both the communication and technical phases of the art.

(d) Expansion of the existing reservoir within the amateur radio service of trained operators, technicians, and electronics experts.

(e) Continuation and extension of the amateur's unique ability to enhance international goodwill.

§97.3 Definitions.

(a) The definitions of terms used in part 97 are:

(1) *Amateur operator.* A person named in an amateur operator/primary license station grant on the ULS consolidated licensee database to be the control operator of an amateur station.

(2) *Amateur radio services.* The amateur service, the amateur-satellite service and the radio amateur civil emergency service.

(4) *Amateur service.* A radiocommunication service for the purpose of self-training, intercommunication and technical investigations carried out by amateurs, that is, duly authorized persons interested in radio technique solely with a personal aim and without pecuniary interest.

(5) *Amateur station.* A station in an amateur radio service consisting of the apparatus necessary for carrying on radiocommunications.

(6) *Automatic control.* The use of devices and procedures for control of a station when it is transmitting so that compliance with the FCC Rules is achieved without the control operator being present at a control point.

(7) *Auxiliary station.* An amateur station, other than in a message forwarding system, that is transmitting communications point-to-point within a system of cooperating amateur stations.

(8) *Bandwidth.* The width of a frequency band outside of which the mean power of the transmitted signal is attenuated at least 26 dB below the mean power of the transmitted signal within the band.

(9) *Beacon.* An amateur station transmitting communications for the purposes of observation of propagation and reception or other related experimental activities.

(10) *Broadcasting.* Transmissions intended for reception by the general public, either direct or relayed.

(11) *Call sign system.* The method used to select a call sign for amateur station over-the-air identification purposes. The call sign systems are:

(i) *Sequential call sign system.* The call sign is selected by the FCC from an alphabetized list corresponding to the geographic region of the licensee's mailing address and operator class. The call sign is shown on the license. The FCC will issue public announcements detailing the procedures of the sequential call sign system.

(ii) *Vanity call sign system.* The call sign is selected by the FCC from a list of call signs requested by the licensee. The call sign is shown on the license. The FCC will issue public announcements detailing the procedures of the vanity call sign system.

(iii) *Special event call sign system.* The call sign is selected by the station licensee from a list of call signs shown on a common data base coordinated, maintained and disseminated by the amateur station special event call sign data base coordinators. The call sign must have the single letter prefix K, N or W, followed by a single numeral Ø through 9, followed by a single letter A through W or Y or Z (for example K1A). The special event call sign is substituted for the call sign shown on the station license grant while the station is transmitting. The FCC will issue public announcements detailing the procedures of the special event call sign system.

(12) *CEPT radio amateur license.* A license issued by a country belonging to the European Conference of Postal and Telecommunications Administrations (CEPT) that has adopted Recommendation T/R 61-01 (Nice 1985, Paris 1992, Nicosia 2003).

(13) *Control operator.* An amateur operator designated by the licensee of a station to be responsible for the transmissions from that station to assure compliance with the FCC Rules.

(14) *Control point.* The location at which the control operator function is performed.

(15) *CSCE.* Certificate of successful completion of an examination.

(16) *Earth station.* An amateur station located on, or within 50 km of, the Earth's surface intended for communications with space stations or with other Earth stations by means of one or more other objects in space.

(17) [Reserved]

(18) *External RF power amplifier.* A device capable of increasing power output when used in conjunction with, but not an integral part of, a transmitter.

(19) [Reserved]

(20) *FAA.* Federal Aviation Administration.

(21) *FCC.* Federal Communications Commission.

(22) *Frequency coordinator.* An entity, recognized in a local or regional area by amateur operators whose stations are eligible to be auxiliary or repeater stations, that recommends transmit/receive channels and associated operating and technical parameters for such stations in order to avoid or minimize potential interference.

(23) *Harmful interference.* Interference which endangers the functioning of a radionavigation service or of other safety services or seriously degrades, obstructs or repeatedly interrupts a radiocommunication service operating in accordance with the Radio Regulations.

(24) *IARP (International Amateur Radio Permit).* A document issued pursuant to the terms of the Inter-American Convention on an International Amateur Radio Permit by a country signatory to that Convention, other than the United States. Montrouis, Haiti. AG/doc.3216/95.

(25) *Indicator.* Words, letters or numerals appended to and separated from the call sign during the station identification.

(26) *Information bulletin.* A message directed only to amateur operators consisting solely of subject matter of direct interest to the amateur service.

(27) *In-law.* A parent, stepparent, sibling, or step-sibling of a licensee's spouse; the spouse of a licensee's sibling, step-sibling, child, or stepchild; or the spouse of a licensee's spouse's sibling or step-sibling.

(28) *International Morse code.* A dot-dash code as defined in ITU-T Recommendation F.1 (March, 1998), Division B, I. Morse code.

(29) *ITU.* International Telecommunication Union.

(30) *Line A.* Begins at Aberdeen, WA, running by great circle arc to the intersection of 48° N, 120° W, thence along parallel 48° N, to the intersection of 95° W, thence by great circle arc through the southernmost point of Duluth, MN, thence by great circle arc to 45° N, 85° W, thence southward along meridian 85° W, to its intersection with parallel 41° N, thence along parallel 41° N, to its intersection with meridian 82° W, thence by great circle arc through the southernmost point of Bangor, ME, thence by great circle arc through the southernmost point of Searsport, ME, at which point it terminates.

(31) *Local control.* The use of a control operator who directly manipulates the operating adjustments in the station to achieve compliance with the FCC Rules.

(32) *Message forwarding system.* A group of amateur stations participating in a voluntary, cooperative, interactive arrangement where communications are sent from the control operator of an originating station to the control operator of one or more destination stations by one or more forwarding stations.

(33) *National Radio Quiet Zone.* The area in Maryland, Virginia and West Virginia Bounded by 39°15′ N on the north, 78°30′ W on the east, 37°30′ N on the south and 80°30′ W on the west.

(34) *Physician.* For the purpose of this part, a person who is licensed to practice in a place where the amateur service is regulated by the FCC, as either a Doctor of Medicine (M.D.) or a Doctor of Osteopathy (D.O.)

(35) *Question pool.* All current examination questions for a designated written examination element.

(36) *Question set.* A series of examination questions on a given examination selected from the question pool.

(37) *Radio Regulations.* The latest ITU *Radio Regulations* to which the United States is a party.

(38) *RACES* (radio amateur civil emergency service). A radio service using amateur stations for civil defense communications during periods of local, regional or national civil emergencies.

(39) *Remote control.* The use of a control operator who indirectly manipulates the operating adjustments in the station through a control link to achieve compliance with the FCC Rules.

(40) *Repeater.* An amateur station that simultaneously retransmits the transmission of another amateur station on a different channel or channels.

(41) *Space station.* An amateur station located more than 50 km above the Earth's surface.

(42) *Space telemetry.* A one-way transmission from a space station of measurements made from the measuring instruments in a spacecraft, including those relating to the functioning of the spacecraft.

(43) *Spurious emission.* An emission, or frequencies outside the necessary bandwidth of a transmission, the level of which may be reduced without affecting the information being transmitted.

(44) *Telecommand.* A one-way transmission to initiate, modify, or terminate functions of a device at a distance.

(45) *Telecommand station.* An amateur station that transmits communications to initiate, modify or terminate functions of a space station.

(46) *Telemetry.* A one-way transmission of measurements at a distance from the measuring instrument.

(47) *Third party communications.* A message from the control operator (first party) of an amateur station to another amateur station control operator (second party) on behalf of another person (third party).

(48) *ULS (Universal Licensing System).* The consolidated database, application filing system and processing system for all Wireless Telecommunications Services.

(49) *VE.* Volunteer examiner.

(50) *VEC.* Volunteer-examiner coordinator.

(b) The definitions of technical symbols used in this part are:

(1) *EHF* (extremely high frequency). The frequency range 30 - 300 GHz.

(2) *EIRP* (equivalent isotropically radiated power). The product of the power supplied to the antenna and the antenna gain in a given direction relative to an isotropic antenna (absolute or isotropic gain).

NOTE: Divide EIRP by 1.64 to convert to effective radiated power.

(3) *ERP* (effective radiated power) (in a given direction). The product of the power supplied to the antenna and its gain relative to a half-wave dipole in a given direction.

NOTE: Multiply ERP by 1.64 to convert to equivalent isotropically radiated power.

(4) *HF* (high frequency). The frequency range 3- 30 MHz.

(5) *Hz.* Hertz.

(6) *LF* (low frequency). The frequency range 30 - 300 kHz.

(7) *m.* Meters.

(8) *MF* (medium frequency). The frequency range 300 - 3000 kHz.

(9) *PEP* (peak envelope power). The average power supplied to the antenna transmission line by a transmitter during one RF cycle at the crest of the modulation envelope taken under normal operating conditions.

(10) *RF.* Radio frequency.

(11) *SHF* (super high frequency). The frequency range 3- 30 GHz.

(12) *UHF* (ultra high frequency). The frequency range 300 - 3000 MHz.

(13) *VHF* (very high frequency). The frequency range 30 - 300 MHz.

(14) *W.* Watts.

(c) The following terms are used in this part to indicate emission types. Refer to §2.201 of the FCC Rules, *Emission, modulation and transmission characteristics,* for information on emission type designators.

(1) *CW.* International Morse code telegraphy emissions having designators with A, C, H, J or R as the first symbol; 1 as the second symbol; A or B as the third symbol; and emissions J2A and J2B.

(2) *Data.* Telemetry, telecommand and computer communications emissions having (i) designators with A, C, D, F, G, H, J or R as the first symbol, 1 as the second symbol, and D as the third symbol; (ii) emission J2D; and (iii) emissions A1C, F1C, F2C, J2C, and J3C having an occupied bandwidth of 500 Hz or less when transmitted on an amateur service frequency below 30 MHz. Only a digital code of a type specifically authorized in this part may be transmitted.

(3) *Image.* Facsimile and television emissions having designators with A, C, D, F, G, H, J or R as the first symbol; 1, 2 or 3 as the second symbol; C or F as the third symbol; and emissions having B as the first symbol; 7, 8 or 9 as the second symbol; W as the third symbol.

(4) *MCW.* Tone-modulated international Morse code telegraphy emissions having designators with A, C, D, F, G, H or R as the first symbol; 2 as the second symbol; A or B as the third symbol.

(5) *Phone.* Speech and other sound emissions having designators with A, C, D, F, G, H, J or R as the first symbol; 1, 2, 3 or X as the second symbol; E as the third symbol. Also speech emissions having B or F as the first symbol; 7, 8 or 9 as the second symbol; E as the third symbol. MCW for the purpose of performing the station identification procedure, or for providing telegraphy practice interspersed with speech. Incidental tones for the purpose of selective calling or alerting or to control the level of a demodulated signal may also be considered phone.

(6) *Pulse.* Emissions having designators with K, L, M, P, Q, V or W as the first symbol; 0, 1, 2, 3, 7, 8, 9 or X as the second symbol; A, B, C, D, E, F, N, W or X as the third symbol.

(7) *RTTY.* Narrow-band direct-printing telegraphy emissions having designators with A, C, D, F, G, H, J or R as the first symbol; 1 as the second symbol; B as the third symbol; and emission J2B. Only a digital code of a type specifically authorized in this part may be transmitted.

(8) *SS.* Spread spectrum emissions using bandwidth-expansion modulation emissions having designators with A, C, D, F, G, H, J or R as the first symbol; X as the second symbol; X as the third symbol.

(9) *Test.* Emissions containing no information having the designators with N as the third symbol. Test does not include pulse emissions with no information or modulation unless pulse emissions are also authorized in the frequency band.

[54 FR 25857, June 20, 1989]

EDITORIAL NOTE: For FEDERAL REGISTER citations affecting §97.3, see the List of CFR Sections Affected, which appears in the Finding Aids section of the printed volume and at *www.fdsys.gov.*

§97.5 Station license required.

(a) The station apparatus must be under the physical control of a person named in an amateur station license grant on the ULS consolidated license database or a person authorized for alien reciprocal operation by §97.107 of this part, before the station may transmit on any amateur service frequency from any place that is:

(1) Within 50 km of the Earth's surface and at a place where the amateur service is regulated by the FCC;

(2) Within 50 km of the Earth's surface and aboard any vessel or craft that is documented or registered in the United States; or

(3) More than 50 km above the Earth's surface aboard any craft that is documented or registered in the United States.

(b) The types of station license grants are:

(1) *An operator/primary station license grant.* One, but only one, operator/primary station license grant may be held by any one person. The primary station license is granted together with the amateur operator license. Except for a representative of a foreign government, any person who qualifies by examination is eligible to apply for an operator/primary station license grant.

(2) *A club station license grant.* A club station license grant may be held only by the person who is the license trustee designated by an officer of the club. The trustee must be a person who holds an operator/primary station license grant. The club must be composed of at least four persons and must have a name, a document of organization, management, and a primary purpose devoted to amateur service activities consistent with this part.

(3) *A military recreation station license grant.* A military recreation station license grant may be held only by the person who is the license custodian designated by the official in charge of the United States military recreational premises where the station is situated. The person must not be a representative of a foreign government. The person need not hold an amateur operator license grant.

(c) The person named in the station license grant or who is authorized for alien reciprocal operation by §97.107 of this part may use, in accordance with the applicable rules of this part, the transmitting apparatus under the physical control of the person at places where the amateur service is regulated by the FCC.

(d) A CEPT radio-amateur license is issued to the person by the country of which the person is a citizen. The person must not:

(1) Be a resident alien or citizen of the United States, regardless of any other citizenship also held;

(2) Hold an FCC-issued amateur operator license nor reciprocal permit for alien amateur licensee;

(3) Be a prior amateur service licensee whose FCC-issued license was revoked, suspended for less than the balance of the license term and the suspension is still in effect, suspended for the balance of the license term and relicensing has not taken place, or surrendered for cancellation following notice of revocation, suspension or monetary forfeiture proceedings; or

(4) Be the subject of a cease and desist order that relates to amateur service operation and which is still in effect.

(e) An IARP is issued to the person by the country of which the person is a citizen. The person must not:

(1) Be a resident alien or citizen of the United States, regardless of any other citizenship also held;

(2) Hold an FCC-issued amateur operator license nor reciprocal permit for alien amateur licensee;

(3) Be a prior amateur service licensee whose FCC-issued license was revoked, suspended for less than the balance of the license term and the suspension is still in effect, suspended for the balance of the license term and relicensing has not taken place, or surrendered for cancellation following notice of revocation, suspension or monetary forfeiture proceedings; or

(4) Be the subject of a cease and desist order that relates to amateur service operation and which is still in effect.

[59 FR 54831, Nov. 2, 1994, as amended at 62 FR 17567, Apr. 10, 1997; 63 FR 68977, Dec. 14, 1998; 75 FR 78169, Dec. 15, 2010]

§97.7 Control operator required.

When transmitting, each amateur station must have a control operator. The control operator must be a person:

(a) For whom an amateur operator/primary station license grant appears on the ULS consolidated licensee database, or

(b) Who is authorized for alien reciprocal operation by §97.107 of this part.

[63 FR 68978, Dec. 14, 1998]

§97.9 Operator license grant.

(a) The classes of amateur operator license grants are: Novice, Technician, General, Advanced, and Amateur Extra. The person named in the operator license grant is authorized to be the control operator of an amateur station with the privileges authorized to the operator class specified on the license grant.

(b) The person named in an operator license grant of Novice, Technician, General or Advanced Class, who has properly submitted to the administering VEs a FCC Form 605 document requesting examination for an operator license grant of a higher class, and who holds a CSCE indicating that the person has completed the necessary examinations within the previous 365 days, is authorized to exercise the rights and privileges of the higher operator class until final disposition of the application or until 365 days following the passing of the examination, whichever comes first.

[75 FR 78169, Dec. 15, 2010]

§97.11 Stations aboard ships or aircraft.

(a) The installation and operation of an amateur station on a ship or aircraft must be approved by the master of the ship or pilot in command of the aircraft.

(b) The station must be separate from and independent of all other radio apparatus installed on the ship or aircraft, except a common antenna may be shared with a voluntary ship radio installation. The station's transmissions must not cause interference to any other apparatus installed on the ship or aircraft.

(c) The station must not constitute a hazard to the safety of life or property. For a station aboard an aircraft, the apparatus shall not be operated while the aircraft is operating under Instrument Flight Rules, as defined by the FAA, unless the station has been found to comply with all applicable FAA Rules.

§97.13 Restrictions on station location.

(a) Before placing an amateur station on land of environmental importance or that is significant in American history, architecture or culture, the licensee may be required to take certain actions prescribed by §§1.1305 - 1.1319 of this chapter.

(b) A station within 1600 m (1 mile) of an FCC monitoring facility must protect that facility from harmful interference. Failure to do so could result in imposition of operating restrictions upon the amateur station pursuant to §97.121. Geographical coordinates of the facilities that require protection are listed in §0.121(c) of this chapter.

(c) Before causing or allowing an amateur station to transmit from any place where the operation of the station could cause human exposure to RF electromagnetic field levels in excess of those allowed under §1.1310 of this chapter, the licensee is required to take certain actions.

(1) The licensee shall ensure compliance with the Commission's radio frequency exposure requirements in §§ 1.1307(b), 2.1091, and 2.1093 of this chapter, where applicable. In lieu of evaluation with the general population/uncontrolled exposure limits, amateur

licensees may evaluate their operation with respect to members of his or her immediate household using the occupational/controlled exposure limits in § 1.1310, provided appropriate training and information has been accessed by the amateur licensee and members of his/her household. RF exposure of other nearby persons who are not members of the amateur licensee's household must be evaluated with respect to the general population/uncontrolled exposure limits. Appropriate methodologies and guidance for evaluating amateur radio service operation is described in the *Office of Engineering and Technology (OET) Bulletin 65,* Supplement B.

(2) If the routine environmental evaluation indicates that the RF electromagnetic fields could exceed the limits contained in §1.1310 of this chapter in accessible areas, the licensee must take action to prevent human exposure to such RF electromagnetic fields. Further information on evaluating compliance with these limits can be found in the FCC's OET Bulletin Number 65, "Evaluating Compliance with FCC Guidelines for Human Exposure to Radiofrequency Electromagnetic Fields."

Table 1 to § 1.1310(e)(1) - Limits for Maximum Permissible Exposure (MPE)

Frequency Range (MHz)	Electric field strength (V/m)	Magnetic field strength (A/m)	Power density (mW/cm^2)	Averaging time (minutes)
{i} Limits for Occupational/Controlled Exposure				
0.2 – 3.0	614	1.63	*(100)	≤6
3.0 – 30	1842/f	4.89/f	*(900/f^2)	≤6
30 – 300	61.4	0.163	1.0	≤6
300 – 1,500			f/300	≤6
1,500 – 100,000			5	≤6
(ii) Limits for General Population/Uncontrolled Exposure				
0.3 0 1.34	614	1.63	*(100)	≤30
1.34 – 30	824/f	2.19/f	*(180/f^2)	≤30
30 – 300	27.5	0.073	0.2	≤30
300 – 1,500			f/1500	≤30
1,500 – 100,000			1.0	≤30
F – frequency in MHz. * = Plane-wave equivalent power density.				

[54 FR 25857, June 20, 1989, as amended at 55 FR 20398, May 16, 1990; 61 FR 41019, Aug. 7, 1996; 62 FR 47963, Sept. 12, 1997; 62 FR 49557, Sept. 22, 1997; 62 FR 61448, Nov. 18, 1997; 63 FR 68978, Dec. 14, 1998; 65 FR 6549, Feb. 10, 2000; 80 FR 53752, Sept. 8, 2015; 85 FR 18151, Apr. 1, 2020]§97.15 Station antenna structures.

§ 97.15 Station antenna structures.

(a) Owners of certain antenna structures more than 60.96 meters (200 feet) above ground level at the site or located near or at a public use airport must notify the Federal Aviation Administration and register with the Commission as required by part 17 of this chapter.

(b) Except as otherwise provided herein, a station antenna structure may be erected at heights and dimensions sufficient to accommodate amateur service communications. (State and local regulation of a station antenna structure must not preclude amateur service communications. Rather, it must reasonably accommodate such communications and must constitute the minimum practicable regulation to accomplish the state or local authority's legitimate purpose. *See* PRB-1, 101 FCC 2d 952 (1985) for details.)

(c) Antennas used to transmit in the 2200 m and 630 m bands must not exceed 60 meters in height above ground level.

[64 FR 53242, Oct. 1, 1999, as amended at 82 FR 27214, June 14, 2017]

§97.17 Application for new license grant.

(a) Any qualified person is eligible to apply for a new operator/primary station, club station or military recreation station license grant. No new license grant will be issued for a Novice or Advanced Class operator/primary station.

(b) Each application for a new amateur service license grant must be filed with the FCC as follows:

(1) Each candidate for an amateur radio operator license which requires the applicant to pass one or more examination elements must present the administering VEs with all information required by the rules prior to the examination. The VEs may collect all necessary information in any manner of their choosing, including creating their own forms.

(2) For a new club or military recreation station license grant, each applicant must present all information required by the rules to an amateur radio organization having tax-exempt status under section 501(c)(3) of the Internal Revenue Code of 1986 that provides voluntary, uncompensated and unreimbursed services in providing club and military recreation station call signs ("*Club Station Call Sign Administrator*") who must submit the information to the FCC in an electronic batch file. The Club Station Call Sign Administrator may collect the information required by these rules in any manner of their choosing, including creating their own forms. The Club Station Call Sign Administrator must retain the applicants information for at least 15 months and make it available to the FCC upon request. The FCC will issue public announcements listing the qualified organizations that have completed a pilot autogrant batch filing project and are authorized to serve as a Club Station Call Sign Administrator.

(c) No person shall obtain or attempt to obtain, or assist another person to obtain or attempt to obtain, an amateur service license grant by fraudulent means.

(d) One unique call sign will be shown on the license grant of each new primary, club and military recreation station. The call sign will be selected by the sequential call sign system. Effective February 14, 2011, no club station license grants will be issued to a licensee who is shown as the license trustee on an existing club station license grant.

[63 FR 68978, Dec. 14, 1998, as amended at 64 FR 53242, Oct. 1, 1999; 65 FR 6549, Feb. 10, 2000; 75 FR 78170, Dec. 15, 2010]

§97.19 Application for a vanity call sign.

(a) The person named in an operator/primary station license grant or in a club station license grant is eligible to make application for modification of the license grant, or the renewal thereof, to show a call sign selected by the vanity call sign system. Effective February 14, 2011, the person named in a club station license grant that shows on the license a call sign that was selected by a trustee is not eligible for an additional vanity call sign. (The person named in a club station license grant that shows on the license a call sign that was selected by a trustee is eligible for a vanity call sign for his or her operator/primary station license grant on the same basis as any other person who holds an operator/primary station license grant.) Military recreation stations are not eligible for a vanity call sign.

(b) Each application for a modification of an operator/primary or club station license grant, or the renewal thereof, to show a call sign selected by the vanity call sign system must be filed in accordance with §1.913 of this chapter.

(c) Unassigned call signs are available to the vanity call sign system with the following exceptions:

(1) A call sign shown on an expired license grant is not available to the vanity call sign system for 2 years following the expiration of the license.

(2) A call sign shown on a surrendered or canceled license grant (except for a license grant that is canceled pursuant to §97.31) is not available to the vanity call sign system for 2 years following the date such action is taken. (The availability of a call sign shown on a license canceled pursuant to §97.31 is governed by paragraph (c)(3) of this section.)

(i) This 2-year period does not apply to any license grant pursuant to paragraph (c)(3)(i), (ii), or (iii) of this section that is surrendered, canceled, revoked, voided, or set aside because the grantee acknowledged, or the Commission determined that the grantee was not eligible for the exception. In such a case, the call sign is not available to the vanity call sign system for 30 days following the date such action is taken, or for the period for which the call sign would not have been available to the vanity call sign system pursuant to paragraphs (c)(2) or (3) of this section but for the intervening grant to the ineligible applicant, whichever is later.

(ii) An applicant to whose operator/primary station license grant, or club station license grant for which the applicant is the trustee, the call sign was previously assigned is exempt from the 2-year period set forth in paragraph (c)(2) of this section.

(3) A call sign shown on a license canceled pursuant to §97.31 of this part is not available to the vanity call sign system for 2 years following the person's death, or for 2 years following the expiration of the license grant, whichever is sooner. If, however, a license is canceled more than 2 years after the licensee's death (or within 30 days before the second anniversary of the licensee's death), the call sign is not available to the vanity call sign system for 30 days following the date such action is taken. The following applicants are exempt from this 2-year period:

(i) An applicant to whose operator/primary station license grant, or club station license grant for which the applicant is the trustee, the call sign was previously assigned; or

(ii) An applicant who is the spouse, child, grandchild, stepchild, parent, grandparent, stepparent, brother, sister, stepbrother, stepsister, aunt, uncle, niece, nephew, or in-law of the person now deceased or of any other deceased former holder of the call sign, provided that the vanity call sign requested by the applicant is from the group of call signs corresponding to the same or lower class of operator license held by the applicant as designated in the sequential call sign system; or

(iii) An applicant who is a club station license trustee acting with a written statement of consent signed by either the licensee *ante mortem* but who is now deceased, or by at least one relative as listed in paragraph (c)(3)(ii) of this section, of the person now deceased or of any other deceased former holder of the call sign, provided that the deceased former holder was a member of the club during his or her life.

(d) The vanity call sign requested by an applicant must be selected from the group of call signs corresponding to the same or lower class of operator license held by the applicant as designated in the sequential call sign system.

(1) The applicant must request that the call sign shown on the license grant be vacated and provide a list of up to 25 call signs in order of preference. In the event that the Commission receives more than one application requesting a vanity call sign from an applicant on the same receipt day, the Commission will process only the first such application entered into the Universal Licensing System. Subsequent vanity call sign applications from that applicant with the same receipt date will not be accepted.

(2) The first assignable call sign from the applicant's list will be shown on the license grant. When none of those call signs are assignable, the call sign vacated by the applicant will be shown on the license grant.

(3) Vanity call signs will be selected from those call signs assignable at the time the application is processed by the FCC.

(4) A call sign designated under the sequential call sign system for Alaska, Hawaii, Caribbean Insular Areas, and Pacific Insular areas will be assigned only to a primary or club station whose licensee's mailing address is in the corresponding state, commonwealth, or island. This limitation does not apply to an applicant for the call sign as the spouse, child, grandchild, stepchild, parent, grandparent, stepparent, brother, sister, stepbrother, stepsister, aunt, uncle, niece, nephew, or in-law, of the former holder now deceased.

[60 FR 7460, Feb. 8, 1995, as amended at 60 FR 50123, Sept. 28, 1995; 60 FR 53132, Oct. 12, 1995; 63 FR 68979, Dec. 14, 1998; 71 FR 66461, Nov. 15, 2006; 75 FR 78170, Dec. 15, 2010]

§97.21 Application for a modified or renewed license grant.

A person holding a valid amateur station license grant:

(1) Must apply to the FCC for a modification of the license grant as necessary to show the correct mailing and email address, licensee name, club name, license trustee name, or license custodian name in accordance with § 1.913 of this chapter. For a club or military recreation station license grant, the application must be presented in document form to a Club Station Call Sign Administrator who must submit the information thereon to the FCC in an electronic batch file. The Club Station Call Sign Administrator must retain the collected information for at least 15 months and make it available to the FCC upon request. A Club Station Call Sign Administrator shall not file with the Commission any application to modify a club station license grant that was submitted by a person other than the trustee as shown on the license grant, except an application to change the club station license trustee. An application to modify a club station license grant to change the license trustee name must be submitted to a Club Station Call Sign Administrator and must be signed by an officer of the club.

(2) May apply to the FCC for a modification of the operator/primary station license grant to show a higher operator class. Applicants must present the administering VEs with all information required by the rules prior to the examination. The VEs may collect all necessary information in any manner of their choosing, including creating their own forms.

(3) May apply to the FCC for renewal of the license grant for another term in accordance with §§ 1.913 and 1.949 of this chapter. Application for renewal of a Technician Plus Class operator/primary station license will be processed as an application for renewal of a Technician Class operator/primary station license.

(i) For a station license grant showing a call sign obtained through the vanity call sign system, the application must be filed in accordance with § 97.19 of this part in order to have the vanity call sign reassigned to the station.

(ii) For a primary station license grant showing a call sign obtained through the sequential call sign system, and for a primary station license grant showing a call sign obtained through the vanity call sign system but whose grantee does not want to have the vanity call sign reassigned to the station, the application must be filed with the FCC in accordance with § 1.913 of this chapter. When the application has been received by the FCC on or before the license expiration date, the license operating authority is continued until the final disposition of the application.

(iii) For a club station or military recreation station license grant showing a call sign obtained through the sequential call sign system, and for a club station license grant showing a call sign obtained through the vanity call sign system but whose grantee does not want to have the vanity call sign reassigned to the station, the application must be presented in document form to a Club Station Call Sign Administrator who must submit the information thereon to the FCC in an electronic batch file. The replacement call sign will be selected by the sequential call sign system. The Club Station Call Sign Administrator must retain the collected information for at least 15 months and make it available to the FCC upon request.

(b) A person whose amateur station license grant has expired may apply to the FCC for renewal of the license grant for another term during a 2 year filing grace period. The application must be received at the address specified above prior to the end of the grace period. Unless and until the license grant is renewed, no privileges in this part are conferred.

(c) Except as provided in paragraph (a)(3) of this section, a call sign obtained under the sequential or vanity call sign system will be reassigned to the station upon renewal or modification of a station license.

[63 FR 68979, Dec. 14, 1998, as amended at 64 FR 53242, Oct. 1, 1999; 65 FR 6550, Feb. 10, 2000; 75 FR 78170, Dec. 15, 2010; 79 FR 35291, July 21, 2014; 85 FR 85532, Dec. 29, 2020]

§97.23 Mailing address.

Each license grant must show the grantee's correct name, mailing address, and email address. The email address must be an address where the grantee can receive electronic correspondence. Revocation of the station license or suspension of the operator license may result when correspondence from the FCC is returned as undeliverable because the grantee failed to provide the correct email address.

[85 FR 85533, Dec. 29, 2020]

§97.25 License term.

An amateur service license is normally granted for a 10 -year term.

[63 FR 68979, Dec. 14, 1998]

§97.27 FCC modification of station license grant.

(a) The FCC may modify a station license grant, either for a limited time or for the duration of the term thereof, if it determines:

(1) That such action will promote the public interest, convenience, and necessity; or

(2) That such action will promote fuller compliance with the provisions of the Communications Act of 1934, as amended, or of any treaty ratified by the United States.

(b) When the FCC makes such a determination, it will issue an order of modification. The order will not become final until the licensee is notified in writing of the proposed action and the grounds and reasons therefor. The licensee will be given reasonable opportunity of no less than 30 days to protest the modification; except that, where safety of life or property is involved, a shorter period of notice may be provided. Any protest by a licensee of an FCC order of modification will be handled in accordance with the provisions of 47 U.S.C. 316.

[59 FR 54833, Nov. 2, 1994, as amended at 63 FR 68979, Dec. 14, 1998]

§97.29 Replacement license grant document.

Each grantee whose amateur station license grant document is lost, mutilated or destroyed may apply to the FCC for a replacement in accordance with §1.913 of this chapter.

[63 FR 68979, Dec. 14, 1998]

§97.31 Cancellation on account of the licensee's death.

(a) A person may request cancellation of an operator/primary station license grant on account of the licensee's death by submitting a signed request that includes a death certificate, obituary, or Social Security Death Index data that shows the person named in the operator/primary station license grant has died. Such a request may be submitted as a pleading associated with the deceased licensee's license. *See* §1.45 of this chapter. In addition, the Commission may cancel an operator/primary station license grant if it becomes aware of the grantee's death through other means. No action will be taken during the last thirty days of the post-expiration grace period (*see* §97.21(b)) on a request to cancel a license due to the licensee's death.

(b) A license that is canceled due to the licensee's death is canceled as of the date of the licensee's death.

[75 FR 78171, Dec. 15, 2010]

Subpart B—Station Operation Standards

§97.101 General standards.

(a) In all respects not specifically covered by FCC Rules each amateur station must be operated in accordance with good engineering and good amateur practice.

(b) Each station licensee and each control operator must cooperate in selecting transmitting channels and in making the most effective use of the amateur service frequencies. No frequency will be assigned for the exclusive use of any station.

(c) At all times and on all frequencies, each control operator must give priority to stations providing emergency communications, except to stations transmitting communications for training drills and tests in RACES.

(d) No amateur operator shall willfully or maliciously interfere with or cause interference to any radio communication or signal.

§97.103 Station licensee responsibilities.

(a) The station licensee is responsible for the proper operation of the station in accordance with the FCC Rules. When the control operator is a different amateur operator than the station licensee, both persons are equally responsible for proper operation of the station.

(b) The station licensee must designate the station control operator. The FCC will presume that the station licensee is also the control operator, unless documentation to the contrary is in the station records.

(c) The station licensee must make the station and the station records available for inspection upon request by an FCC representative.

[54 FR 25857, June 20, 1989, as amended at 71 FR 66462, Nov. 15, 2006; 75 FR 27201, May 14, 2010]

§97.105 Control operator duties.

(a) The control operator must ensure the immediate proper operation of the station, regardless of the type of control.

(b) A station may only be operated in the manner and to the extent permitted by the privileges authorized for the class of operator license held by the control operator.

§97.107 Reciprocal operating authority.

A non-citizen of the United States ("alien") holding an amateur service authorization granted by the alien's government is authorized to be the control operator of an amateur station located at places where the amateur service is regulated by the FCC, provided there is in effect a multilateral or bilateral reciprocal operating arrangement, to which the United States and the alien's government are parties, for amateur service operation on a reciprocal basis. The FCC will issue public announcements listing the countries with which the United States has such an arrangement. No citizen of the United States or person holding an FCC amateur operator/primary station license grant is eligible for the reciprocal operating authority granted by this section. The privileges granted to a control operator under this authorization are:

(a) For an amateur service license granted by the Government of Canada:

(1) The terms of the *Convention Between the United States and Canada* (TIAS No. 2508) *Relating to the Operation by Citizens of Either Country of Certain Radio Equipment or Stations in the Other Country;*

(2) The operating terms and conditions of the amateur service license issued by the Government of Canada; and

(3) The applicable rules of this part, but not to exceed the control operator privileges of an FCC-granted Amateur Extra Class operator license.

(b) For an amateur service license granted by any country, other than Canada, with which the United States has a multilateral or bilateral agreement:

(1) The terms of the agreement between the alien's government and the United States;

(2) The operating terms and conditions of the amateur service license granted by the alien's government;

(3) The applicable rules of this part, but not to exceed the control operator privileges of an FCC-granted Amateur Extra Class operator license; and

(c) At any time the FCC may, in its discretion, modify, suspend or cancel the reciprocal operating authority granted to any person by this section.

[63 FR 68979, Dec. 14, 1998]

§97.109 Station control.

(a) Each amateur station must have at least one control point.

(b) When a station is being locally controlled, the control operator must be at the control point. Any station may be locally controlled.

(c) When a station is being remotely controlled, the control operator must be at the control point. Any station may be remotely controlled.

(d) When a station is being automatically controlled, the control operator need not be at the control point. Only stations specifically designated elsewhere in this part may be automatically controlled. Automatic control must cease upon notification by a Regional Director that the station is transmitting improperly or causing harmful interference to other stations. Automatic control must not be resumed without prior approval of the Regional Director.

[54 FR 39535, Sept. 27, 1989, as amended at 60 FR 26001, May 16, 1995; 69 FR 24997, May 5, 2004; 80 FR 53753, Sept. 8, 2015]

§97.111 Authorized transmissions.

(a) An amateur station may transmit the following types of two-way communications:

(1) Transmissions necessary to exchange messages with other stations in the amateur service, except those in any country whose administration has notified the ITU that it objects to such communications. The FCC will issue public notices of current arrangements for international communications.

(2) Transmissions necessary to meet essential communication needs and to facilitate relief actions.

(3) Transmissions necessary to exchange messages with a station in another FCC-regulated service while providing emergency communications;

(4) Transmissions necessary to exchange messages with a United States government station, necessary to providing communications in RACES; and

(5) Transmissions necessary to exchange messages with a station in a service not regulated by the FCC but authorized by the FCC to communicate with amateur stations. An amateur station may exchange messages with a participating United States military station during an Armed Forces Day Communications Test.

(b) In addition to one-way transmissions specifically authorized elsewhere in this part, an amateur station may transmit the following types of one-way communications:

(1) Brief transmissions necessary to make adjustments to the station;

(2) Brief transmissions necessary to establishing two-way communications with other stations;

(3) Telecommand;

(4) Transmissions necessary to providing emergency communications;

(5) Transmissions necessary to assisting persons learning, or improving proficiency in, the international Morse code; and

(6) Transmissions necessary to disseminate information bulletins.

(7) Transmissions of telemetry.

[54 FR 25857, June 20, 1989, as amended at 56 FR 56171, Nov. 1, 1991; 71 FR 25982, May 3, 2006; 71 FR 66462, Nov. 15, 2006]

§97.113 Prohibited transmissions.

(a) No amateur station shall transmit:

(1) Communications specifically prohibited elsewhere in this part;

(2) Communications for hire or for material compensation, direct or indirect, paid or promised, except as otherwise provided in these rules;

(3) Communications in which the station licensee or control operator has a pecuniary interest, including communications on behalf of an employer, with the following exceptions:

(i) A station licensee or station control operator may participate on behalf of an employer in an emergency preparedness or disaster readiness test or drill, limited to the duration and scope of such test or drill, and operational testing immediately prior to such test or drill. Tests or drills that are not government-sponsored are limited to a total time of one hour per week; except that no more than twice in any calendar year, they may be conducted for a period not to exceed 72 hours.

(ii) An amateur operator may notify other amateur operators of the availability for sale or trade of apparatus normally used in an amateur station, provided that such activity is not conducted on a regular basis.

(iii) A control operator may accept compensation as an incident of a teaching position during periods of time when an amateur station is used by that teacher as a part of classroom instruction at an educational institution.

(iv) The control operator of a club station may accept compensation for the periods of time when the station is transmitting telegraphy practice or information bulletins, provided that the station transmits such telegraphy practice and bulletins for at least 40 hours per week; schedules operations on at least six amateur service MF and HF bands using reasonable measures to maximize coverage; where the schedule of normal operating times and frequencies is published at least 30 days in advance of the actual transmissions; and where the control operator does not accept any direct or indirect compensation for any other service as a control operator.

(4) Music using a phone emission except as specifically provided elsewhere in this section; communications intended to facilitate a criminal act; messages encoded for the purpose of obscuring their meaning, except as otherwise provided herein; obscene or indecent words or language; or false or deceptive messages, signals or identification.

(5) Communications, on a regular basis, which could reasonably be furnished alternatively through other radio services.

(b) An amateur station shall not engage in any form of broadcasting, nor may an amateur station transmit one-way communications except as specifically provided in these rules; nor shall an amateur station engage in any activity related to program production or news gathering for broadcasting purposes, except that communications directly related to the immediate safety of human life or the protection of property may be provided by amateur stations to broadcasters for dissemination to the public where no other means of communication is reasonably available before or at the time of the event.

(c) No station shall retransmit programs or signals emanating from any type of radio station other than an amateur station, except propagation and weather forecast information intended for use by the general public and originated from United States Government stations, and communications, including incidental music, originating on United States Government frequencies between a manned spacecraft and its associated Earth stations. Prior approval for manned spacecraft communications retransmissions must be obtained from the National Aeronautics and Space Administration. Such retransmissions must be for the exclusive use of amateur radio operators. Propagation, weather forecasts, and manned spacecraft communications retransmissions may not be conducted on a regular basis, but only occasionally, as an incident of normal amateur radio communications.

(d) No amateur station, except an auxiliary, repeater, or space station, may automatically retransmit the radio signals of other amateur station.

[58 FR 43072, Aug. 13, 1993; 58 FR 47219, Sept. 8, 1993, as amended at 71 FR 25982, May 3, 2006; 71 FR 66462, Nov. 15, 2006; 75 FR 46857, Aug. 4, 2010; 79 FR 35291, June 20, 2014]

§97.115 Third party communications.

(a) An amateur station may transmit messages for a third party to:

(1) Any station within the jurisdiction of the United States.

(2) Any station within the jurisdiction of any foreign government when transmitting emergency or disaster relief communications and any station within the jurisdiction of any foreign government whose administration has made arrangements with the United States to allow amateur stations to be used for transmitting international communications on behalf of third parties. No station shall transmit messages for a third party to any station within the jurisdiction of any foreign government whose administration has not made such an arrangement. This prohibition does not apply to a message for any third party who is eligible to be a control operator of the station.

(b) The third party may participate in stating the message where:

(1) The control operator is present at the control point and is continuously monitoring and supervising the third party's participation; and

(2) The third party is not a prior amateur service licensee whose license was revoked or not renewed after hearing and re-licensing has not taken place; suspended for less than the balance of the license term and the suspension is still in effect; suspended for the balance of the license term and re-licensing has not taken place; or surrendered for cancellation following notice of revocation, suspension or monetary forfeiture proceedings. The third party may not be the subject of a cease and desist order which relates to amateur service operation and which is still in effect.

(c) No station may transmit third party communications while being automatically controlled except a station transmitting a RTTY or data emission.

(d) At the end of an exchange of international third party communications, the station must also transmit in the station identification procedure the call sign of the station with which a third party message was exchanged.

[54 FR 25857, June 20, 1989; 54 FR 39535, Sept. 27, 1989, as amended at 71 FR 25982, May 3, 2006; 71 FR 66462, Nov. 15, 2006]

§97.117 International communications.

Transmissions to a different country, where permitted, shall be limited to communications incidental to the purposes of the amateur service and to remarks of a personal character.

[71 FR 25982, May 3, 2006]

§97.119 Station identification.

(a) Each amateur station, except a space station or telecommand station, must transmit its assigned call sign on its transmitting channel at the end of each communication, and at least every 10 minutes during a communication, for the purpose of clearly making the source of the transmissions from the station known to those receiving the transmissions. No station may transmit unidentified communications or signals, or transmit as the station call sign, any call sign not authorized to the station.

(b) The call sign must be transmitted with an emission authorized for the transmitting channel in one of the following ways:

(1) By a CW emission. When keyed by an automatic device used only for identification, the speed must not exceed 20 words per minute;

(2) By a phone emission in the English language. Use of a phonetic alphabet as an aid for correct station identification is encouraged;

(3) By a RTTY emission using a specified digital code when all or part of the communications are transmitted by a RTTY or data emission;

(4) By an image emission conforming to the applicable transmission standards, either color or monochrome, of §73.682(a) of the FCC Rules when all or part of the communications are transmitted in the same image emission

(c) One or more indicators may be included with the call sign. Each indicator must be separated from the call sign by the slant mark (/) or by any suitable word that denotes the slant mark. If an indicator is self-assigned, it must be included before, after, or both before and after, the call sign. No self-assigned indicator may conflict with any other indicator specified by the FCC Rules or with any prefix assigned to another country.

(d) When transmitting in conjunction with an event of special significance, a station may substitute for its assigned call sign a special event call sign as shown for that station for that period of time on the common data base coordinated, maintained and disseminated by the special event call sign data base coordinators. Additionally, the station must transmit its assigned call sign at least once per hour during such transmissions.

(e) When the operator license class held by the control operator exceeds that of the station licensee, an indicator consisting of the call sign assigned to the control operator's station must be included after the call sign.

(f) When the control operator is a person who is exercising the rights and privileges authorized by §97.9(b) of this part, an indicator must be included after the call sign as follows:

(1) For a control operator who has requested a license modification from Novice Class to Technical Class: KT;

(2) For a control operator who has requested a license modification from Novice or Technician to General Class: AG;

(3) For a control operator who has requested a license modification from Novice, Technician, General, or Advanced Class to Amateur Extra Class: AE.

(g) When the station is transmitting under the authority of §97.107 of this part, an indicator consisting of the appropriate letter-numeral designating the station location must be included before the call sign that was issued to the station by the country granting the license. For an amateur service license granted by the Government of Canada, however, the indicator must be included after the call sign. At least once during each intercommunication, the identification announcement must include the geographical location as nearly as possible by city and state, commonwealth or possession.

[54 FR 25857, June 20, 1989, as amended at 54 FR 39535, Sept. 27, 1989; 55 FR 30457, July 26, 1990; 56 FR 28, Jan. 2, 1991; 62 FR 17567, Apr. 10, 1997; 63 FR 68980, Dec. 14, 1998; 64 FR 51471, Sept. 23, 1999; 66 FR 20752, Apr. 25, 2001; 75 FR 78171, Dec. 15, 2010]

§97.121 Restricted operation.

(a) If the operation of an amateur station causes general interference to the reception of transmissions from stations operating in the domestic broadcast service when receivers of good engineering design, including adequate selectivity characteristics, are used to receive such transmissions, and this fact is made known to the amateur station licensee, the amateur station shall not be operated during the hours from 8 p.m. to 10:30 p.m., local time, and on Sunday for the additional period from 10:30 a.m. until 1 p.m., local time, upon the frequency or frequencies used when the interference is created.

(b) In general, such steps as may be necessary to minimize interference to stations operating in other services may be required after investigation by the FCC.

Subpart C—Special Operations

§97.201 Auxiliary station.

(a) Any amateur station licensed to a holder of a Technician, General, Advanced or Amateur Extra Class operator license may be an auxiliary station. A holder of a Technician, General, Advanced or Amateur Extra Class operator license may be the control operator of an auxiliary station, subject to the privileges of the class of operator license held.

(b) An auxiliary station may transmit only on the 2 m and shorter wavelength bands, except the 144.0 - 144.5 MHz, 145.8 - 146.0 MHz, 219 - 220 MHz, 222.00 - 222.15 MHz, 431 - 433 MHz, and 435 - 438 MHz segments.

(c) Where an auxiliary station causes harmful interference to another auxiliary station, the licensees are equally and fully responsible for resolving the interference unless one station's operation is recommended by a frequency coordinator and the other station's is not. In that case, the licensee of the non-coordinated auxiliary station has primary responsibility to resolve the interference.

(d) An auxiliary station may be automatically controlled.

(e) An auxiliary station may transmit one-way communications.

[54 FR 25857, June 20, 1989, as amended at 56 FR 56171, Nov. 1, 1991; 60 FR 15687, Mar. 27, 1995; 63 FR 68980, Dec. 14, 1998; 71 FR 66462, Nov. 15, 2006; 75 FR 78171, Dec. 15, 2010]

§97.203 Beacon station.

(a) Any amateur station licensed to a holder of a Technician, General, Advanced or Amateur Extra Class operator license may be a beacon. A holder of a Technician, General, Advanced or Amateur Extra Class operator license may be the control operator of a beacon, subject to the privileges of the class of operator license held.

(b) A beacon must not concurrently transmit on more than 1 channel in the same amateur service frequency band, from the same station location.

(c) The transmitter power of a beacon must not exceed 100 W.

(d) A beacon may be automatically controlled while it is transmitting on the 28.20 - 28.30 MHz, 50.06 - 50.08 MHz, 144.275 - 144.300 MHz, 222.05 - 222.06 MHz or 432.300 - 432.400 MHz segments, or on the 33 cm and shorter wavelength bands.

(e) Before establishing an automatically controlled beacon in the National Radio Quiet Zone or before changing the transmitting frequency, transmitter power, antenna height or directivity, the station licensee must give written notification thereof to the Interference Office, National Radio Astronomy Observatory, P.O. Box 2, Green Bank, WV 24944.

(1) The notification must include the geographical coordinates of the antenna, antenna ground elevation above mean sea level (AMSL), antenna center of radiation above ground level (AGL), antenna directivity, proposed frequency, type of emission, and transmitter power.

(2) If an objection to the proposed operation is received by the FCC from the National Radio Astronomy Observatory at Green Bank, Pocahontas County, WV, for itself or on behalf of the Naval Research Laboratory at Sugar Grove, Pendleton County, WV, within 20 days from the date of notification, the FCC will consider all aspects of the problem and take whatever action is deemed appropriate.

(f) A beacon must cease transmissions upon notification by a Regional Director that the station is operating improperly or causing undue interference to other operations. The beacon may not resume transmitting without prior approval of the Regional Director.

(g) A beacon may transmit one-way communications.

[54 FR 25857, June 20, 1989, as amended at 55 FR 9323, Mar. 13, 1990; 56 FR 19610, Apr. 29, 1991; 56 FR 32517, July 17, 1991; 62 FR 55536, Oct. 27, 1997; 63 FR 41204, Aug. 3, 1998; 63 FR 68980, Dec. 14, 1998; 69 FR 24997, May 5, 2004; 71 FR 66462, Nov. 15, 2006; 75 FR 78171, Dec. 15, 2010; 80 FR 53753, Sept. 8, 2015]

§97.205 Repeater station.

(a) Any amateur station licensed to a holder of a Technician, General, Advanced or Amateur Extra Class operator license may be a repeater. A holder of a Technician, General, Advanced or Amateur Extra Class operator license may be the control operator of a repeater, subject to the privileges of the class of operator license held.

(b) A repeater may receive and retransmit only on the 10 m and shorter wavelength frequency bands except the 28.0 - 29.5 MHz, 50.0 - 51.0 MHz, 144.0 - 144.5 MHz, 145.5 - 146.0 MHz, 222.00 - 222.15 MHz, 431.0 - 433.0 Mhz, and 435.0 - 438.0 Mhz segments.

(c) Where the transmissions of a repeater cause harmful interference to another repeater, the two station licensees are equally and fully responsible for resolving the interference unless the operation of one station is recommended by a frequency coordinator and the operation of the other station is not. In that case, the licensee of the non-coordinated repeater has primary responsibility to resolve the interference.

(d) A repeater may be automatically controlled.

(e) Ancillary functions of a repeater that are available to users on the input channel are not considered remotely controlled functions of the station. Limiting the use of a repeater to only certain user stations is permissible.

(f) [Reserved]

(g) The control operator of a repeater that retransmits inadvertently communications that violate the rules in this part is not accountable for the violative communications.

(h) The provisions of this paragraph do not apply to repeaters that transmit on the 1.2 cm or shorter wavelength bands. Before establishing a repeater within 16 km (10 miles) of the Arecibo Observatory or before changing the transmitting frequency, transmitter power, antenna height or directivity of an existing repeater, the station licensee must give written notification thereof to the Interference Office, Arecibo Observatory, HC3 Box 53995, Arecibo, Puerto Rico 00612, in writing or electronically, of the technical parameters of the proposal. Licensees who choose to transmit information electronically should e-mail to: *prcz@naic.edu*.

(1) The notification shall state the geographical coordinates of the antenna (NAD- 83 datum), antenna height above mean sea level (AMSL), antenna center of radiation above ground level (AGL), antenna directivity and gain, proposed frequency and FCC Rule Part, type of emission, effective radiated power, and whether the proposed use is itinerant. Licensees may wish to consult interference guidelines provided by Cornell University.

(2) If an objection to the proposed operation is received by the FCC from the Arecibo Observatory, Arecibo, Puerto Rico, within 20 days from the date of notification, the FCC will consider all aspects of the problem and take whatever action is deemed appropriate. The licensee will be required to make reasonable efforts in order to resolve or mitigate any potential interference problem with the Arecibo Observatory.

[54 FR 25857, June 20, 1989, as amended at 55 FR 4613, Feb. 9, 1990; 56 FR 32517, July 17, 1991; 58 FR 64385, Dec. 7, 1993; 59 FR 18975, Apr. 21, 1994; 62 FR 55536, Oct. 27, 1997; 63 FR 41205, Aug. 3, 1998; 63 FR 68980, Dec. 14, 1998; 69 FR 24997, May 5, 2004; 70 FR 31374, June 1, 2005]

§97.207 Space station.

(a) Any amateur station may be a space station. A holder of any class operator license may be the control operator of a space station, subject to the privileges of the class of operator license held by the control operator.

(b) A space station must be capable of effecting a cessation of transmissions by telecommand whenever such cessation is ordered by the FCC.

(c) The following frequency bands and segments are authorized to space stations:

(1) The 17 m, 15 m, 12 m, and 10 m bands, 6 mm, 4 mm, 2 mm and 1 mm bands; and

(2) The 7.0-7.1 MHz, 14.00-14.25 MHz, 144-146 MHz, 435-438 MHz, 2400-2450 MHz, 5.83-5.85 GHz, 10.45-10.50 GHz, and 24.00-24.05 GHz segments.

(d) A space station may automatically retransmit the radio signals of Earth stations and other space stations.

(e) A space station may transmit one-way communications.

(f) Space telemetry transmissions may consist of specially coded messages intended to facilitate communications or related to the function of the spacecraft.

(g) The license grantee of each space station must make the following written notifications to the Space Bureau, FCC, Washington, DC 20554.

(1) A pre-space notification within 30 days after the date of launch vehicle determination, but no later than 90 days before integration of the space station into the launch vehicle. The notification must be in accordance with the provisions of Articles 9 and 11 of the International Telecommunication Union (ITU) Radio Regulations and must specify the information required by Appendix 4 and Resolution No. 642 of the ITU Radio Regulations. The notification must also include a description of the design and operational strategies that the space station will use to mitigate orbital debris, including the following information:

(i) A statement that the space station operator has assessed and limited the amount of debris released in a planned manner during normal operations. Where applicable, this statement must include an orbital debris mitigation disclosure for any separate deployment devices, distinct from the space station launch vehicle, that may become a source of orbital debris;

(ii) A statement indicating whether the space station operator has assessed and limited the probability that the space station(s) will become a source of debris by collision with small debris or meteoroids that would cause loss of control and prevent disposal. The statement must indicate whether this probability for an individual space station is 0.01 (1 in 100) or less, as calculated using the NASA Debris Assessment Software or a higher fidelity assessment tool;

(iii) A statement that the space station operator has assessed and limited the probability, during and after completion of mission operations, of accidental explosions or of release of liquids that will persist in droplet form. This statement must include a demonstration that debris generation will not result from the conversion of energy sources on board the spacecraft into energy that fragments the spacecraft. Energy sources include chemical, pressure, and kinetic energy. This demonstration should address whether stored energy will be removed at the spacecraft's end of life, by depleting residual fuel and leaving all fuel line valves open, venting any pressurized system, leaving all batteries in a permanent discharge state, and removing any remaining source of stored energy, or through other equivalent procedures specifically disclosed in the application;

(iv) A statement that the space station operator has assessed and limited the probability of the space station(s) becoming a source of debris by collisions with large debris or other operational space stations.

(A) Where the application is for an NGSO space station or system, the following information must also be included:

(*1*) A demonstration that the space station operator has assessed and limited the probability of collision between any space station of the system and other large objects (10 cm or larger in diameter) during the total orbital lifetime of the space station, including any de-orbit phases, to less than 0.001 (1 in 1,000). The probability shall be calculated using the NASA Debris Assessment Software or a higher fidelity assessment tool. The collision risk may be assumed zero for a space station during any period in which the space station will be maneuvered effectively to avoid colliding with large objects.

(*2*) The statement must identify characteristics of the space station(s)' orbits that may present a collision risk, including any planned and/or operational space stations in those orbits, and indicate what steps, if any, have been taken to coordinate with the other spacecraft or system, or what other measures the operator plans to use to avoid collision.

(*3*) If at any time during the space station(s)' mission or de-orbit phase the space station(s) will transit through the orbits used by any inhabitable spacecraft, including the International Space Station, the statement must describe the design and operational strategies, if any, that will be used to minimize the risk of collision and avoid posing any operational constraints to the inhabitable spacecraft.

(*4*) The statement must disclose the accuracy, if any, with which orbital parameters will be maintained, including apogee, perigee, inclination, and the right ascension of the ascending node(s). In the event that a system is not be maintained to specific orbital tolerances, *e.g.*, its propulsion system will not be used for orbital maintenance, that fact should be included in the debris mitigation disclosure. Such systems must also indicate the anticipated evolution over time of the orbit of the proposed satellite or satellites. All systems must describe the extent of satellite maneuverability, whether or not the space station design includes a propulsion system.

(*5*) The space station operator must certify that upon receipt of a space situational awareness conjunction warning, the operator will review and take all possible steps to assess the collision risk, and will mitigate the collision risk if necessary. As appropriate, steps to assess and mitigate the collision risk should include, but are not limited to: Contacting the operator of any active spacecraft involved in such a warning; sharing ephemeris data and other appropriate operational information with any such operator; and modifying space station attitude and/or operations.

(B) Where a space station requests the assignment of a geostationary orbit location, it must assess whether there are any known satellites located at, or reasonably expected to be located at, the requested orbital location, or assigned in the vicinity of that location, such that the station keeping volumes of the respective satellites might overlap or touch. If so, the statement must include a statement as to the identities of those parties and the measures that will be taken to prevent collisions.

(v) A statement addressing the trackability of the space station(s). Space station(s) operating in low-Earth orbit will be presumed trackable if each individual space station is 10 cm or larger in its smallest dimension, exclusive of deployable components. Where the application is for an NGSO space station or system, the statement shall also disclose the following:

(A) How the operator plans to identify the space station(s) following deployment and whether space station tracking will be active or passive;

(B) Whether, prior to deployment, the space station(s) will be registered with the 18th Space Control Squadron or successor entity; and

(C) The extent to which the space station operator plans to share information regarding initial deployment, ephemeris, and/or planned maneuvers with the 18th Space Control Squadron or successor entity, other entities that engage in space situational awareness or space traffic management functions, and/or other operators.

(vi) A statement disclosing planned proximity operations, if any, and addressing debris generation that will or may result from the proposed operations, including any planned release of debris, the risk of accidental explosions, the risk of accidental collision, and measures taken to mitigate those risks.

(vii) A statement detailing the disposal plans for the space station, including the quantity of fuel—if any—that will be reserved for disposal maneuvers. In addition, the following specific provisions apply:

(A) For geostationary orbit space stations, the statement must disclose the altitude selected for a disposal orbit and the calculations that are used in deriving the disposal altitude.

(B) For space stations terminating operations in an orbit in or passing through the low-Earth orbit region below 2,000 km altitude, the statement must disclose whether the spacecraft will be disposed of either through atmospheric re-entry, specifying if direct retrieval of the spacecraft will be used. The statement must also disclose the expected time in orbit for the space station following the completion of the mission.

(C) For space stations not covered by either paragraph (g)(1)(vii)(A) or (B) of this section, the statement must indicate whether disposal will involve use of a storage orbit or long-term atmospheric re-entry and rationale for the selected disposal plan.

(D) For all NGSO space stations under paragraph (g)(1)(vii)(B) or (C) of this section, the following additional specific provisions apply:

(*1*) The statement must include a demonstration that the probability of success of the chosen disposal method will be 0.9 or greater for any individual space station. For space station systems consisting of multiple space stations, the demonstration should include additional information regarding efforts to achieve a higher probability of success, with a goal, for large systems, of a probability of success for any individual space station of 0.99 or better. For space stations under paragraph (g)(1)(vii)(B) of this section that will be terminating operations in or passing through low-Earth orbit, successful disposal is defined as atmospheric re-entry of the spacecraft within 25 years or less following completion of the mission. For space stations under paragraph (g)(1)(vii)(C) of this section, successful disposal will be assessed on a case-by-case basis.

(*2*) If planned disposal is by atmospheric re-entry, the statement must also include:

(*i*) A disclosure indicating whether the atmospheric re-entry will be an uncontrolled re-entry or a controlled targeted reentry.

(*ii*) An assessment as to whether portions of any individual spacecraft will survive atmospheric re-entry and impact the surface of the Earth with a kinetic energy in excess of 15 joules, and demonstration that the calculated casualty risk for an individual spacecraft using the NASA Debris Assessment Software or a higher fidelity assessment tool is less than 0.0001 (1 in 10,000).

(viii) If any material item described in this notification changes before launch, a replacement pre-space notification shall be filed with the Space Bureau no later than 90 days before integration of the space station into the launch vehicle.

(2) An in-space station notification is required no later than 7 days following initiation of space station transmissions. This notification must update the information contained in the pre-space notification.

(3) A post-space station notification is required no later than 3 months after termination of the space station transmissions. When termination of transmissions is ordered by the FCC, the notification is required no later than 24 hours after termination of transmissions.

[54 FR 25857, June 20, 1989, as amended at 54 FR 39535, Sept. 27, 1989; 56 FR 56171, Nov. 1, 1991; 57 FR 32736, July 23, 1992; 60 FR 50124, Sept. 28, 1995; 63 FR 68980, Dec. 14, 1998; 69 FR 54588, Sept. 9, 2004; 71 FR 66462, Nov. 15, 2006; 75 FR 27201, May 14, 2010; 85 FR 52453, Aug. 25, 2020; 85 FR 64068, Oct. 9, 2020; 88 FR 21451, Apr. 10, 2023]

§97.209 Earth station.

(a) Any amateur station may be an Earth station. A holder of any class operator license may be the control operator of an Earth station, subject to the privileges of the class of operator license held by the control operator.

(b) The following frequency bands and segments are authorized to Earth stations:

(1) The 17 m, 15 m, 12 m, and 10 m bands, 6 mm, 4 mm, 2 mm and 1 mm bands; and

(2) The 7.0-7.1 MHz, 14.00-14.25 MHz, 144-146 MHz, 435-438 MHz, 1260-1270 MHz and 2400-2450 MHz, 5.65-5.67 GHz, 10.45-10.50 GHz and 24.00-24.05 GHz segments.

[54 FR 25857, June 20, 1989, as amended at 54 FR 39535, Sept. 27, 1989; 85 FR 64068, Oct. 9, 2020; 85 FR 69515, Nov. 3, 2020]

§97.211 Space telecommand station.

(a) Any amateur station designated by the licensee of a space station is eligible to transmit as a telecommand station for that space station, subject to the privileges of the class of operator license held by the control operator.

(b) A telecommand station may transmit special codes intended to obscure the meaning of telecommand messages to the station in space operation.

(c) The following frequency bands and segments are authorized to telecommand stations:

(1) The 17 m, 15 m, 12 m and 10 m bands, 6 mm, 4 mm, 2 mm and 1 mm bands; and

(2) The 7.0-7.1 MHz, 14.00-14.25 MHz, 144-146 MHz, 435-438 MHz, 1260-1270 MHz and 2400-2450 MHz, 5.65-5.67 GHz, 10.45-10.50 GHz and 24.00-24.05 GHz segments.

(d) A telecommand station may transmit one-way communications.

[54 FR 25857, June 20, 1989, as amended at 54 FR 39535, Sept. 27, 1989; 56 FR 56171, Nov. 1, 1991; 85 FR 64068, Oct. 9, 2020]

§97.213 Telecommand of an amateur station.

An amateur station on or within 50 km of the Earth's surface may be under telecommand where:

(a) There is a radio or wireline control link between the control point and the station sufficient for the control operator to perform his/her duties. If radio, the control link must use an auxiliary station. A control link using a fiber optic cable or another telecommunication service is considered wireline.

(b) Provisions are incorporated to limit transmission by the station to a period of no more than 3 minutes in the event of malfunction in the control link.

(c) The station is protected against making, willfully or negligently, unauthorized transmissions.

(d) A photocopy of the station license and a label with the name, address, and telephone number of the station licensee and at least one designated control operator is posted in a conspicuous place at the station location.

[54 FR 25857, June 20, 1989, as amended at 56 FR 56171, Nov. 1, 1991]

§97.215 Telecommand of model craft.

An amateur station transmitting signals to control a model craft may be operated as follows:

(a) The station identification procedure is not required for transmissions directed only to the model craft, provided that a label indicating the station call sign and the station licensee's name and address is affixed to the station transmitter.

(b) The control signals are not considered codes or ciphers intended to obscure the meaning of the communication.

(c) The transmitter power must not exceed 1 W.

[54 FR 25857, June 20, 1989, as amended at 56 FR 56171, Nov. 1, 1991]

§97.217 Telemetry.

Telemetry transmitted by an amateur station on or within 50 km of the Earth's surface is not considered to be codes or ciphers intended to obscure the meaning of communications.

[56 FR 56172, Nov. 1, 1991. Redesignated at 59 FR 18975, Apr. 21, 1994]

§97.219 Message forwarding system.

(a) Any amateur station may participate in a message forwarding system, subject to the privileges of the class of operator license held.

(b) For stations participating in a message forwarding system, the control operator of the station originating a message is primarily accountable for any violation of the rules in this part contained in the message.

(c) Except as noted in (d) of this section, for stations participating in a message forwarding system, the control operators of forwarding stations that retransmit inadvertently communications that violate the rules in this part are not accountable for the violative communications. They are, however, responsible for discontinuing such communications once they become aware of their presence.

(d) For stations participating in a message forwarding system, the control operator of the first forwarding station must:

(1) Authenticate the identity of the station from which it accepts communications on behalf of the system; or

(2) Accept accountability for any violation of the rules in this part contained in messages it retransmits to the system.

[59 FR 18975, Apr. 21, 1994]

§97.221 Automatically controlled digital station.

(a) This rule section does not apply to an auxiliary station, a beacon station, a repeater station, an earth station, a space station, or a space telecommand station.

(b) A station may be automatically controlled while transmitting a RTTY or data emission on the 6 m or shorter wavelength bands, and on the 28.120 - 28.189 MHz, 24.925 - 24.930 MHz, 21.090 - 21.100 MHz, 18.105 - 18.110 MHz, 14.0950 - 14.0995 MHz, 14.1005 - 14.112 MHz, 10.140 - 10.150 MHz, 7.100 - 7.105 MHz, or 3.585 - 3.600 MHz segments.

(c) Except for channels specified in §97.303(h), a station may be automatically controlled while transmitting a RTTY or data emission on any other frequency authorized for such emission types provided that:

(1) The station is responding to interrogation by a station under local or remote control; and

(2) No transmission from the automatically controlled station occupies a bandwidth of more than 500 Hz.

[60 FR 26001, May 16, 1995, as amended at 72 FR 3082, Jan. 24, 2007; 77 FR 5412, Feb. 3, 2012]

Subpart D—Technical Standards

§97.301 Authorized frequency bands.

The following transmitting frequency bands are available to an amateur station located within 50 km of the Earth's surface, within the specified ITU Region, and outside any area where the amateur service is regulated by any authority other than the FCC.

(a) For a station having a control operator who has been granted a Technician, General, Advanced, or Amateur Extra Class operator license or who holds a CEPT radio-amateur license or IARP of any class:

Wavelength band	ITU region 1	ITU region 2	ITU region 3	Sharing requirements see §97.303 (paragraph)
VHF	MHz	MHz	MHz	
6 m		50 - 54	50 - 54	(a)
2 m	144 - 146	144 - 148	144 - 148	(a), (k)
1.25 m		219 - 220		(l)
1.25 m		222 - 225		(a)
UHF	MHz	MHz	MHz	
70 cm	430 - 440	420 - 450	430 - 440	(a), (b), (m)
33 cm		902 - 928		(a), (b), (e), (n)
23 cm	1240 - 1300	1240 - 1300	1240 - 1300	(b), (d), (o)
13 cm	2300 - 2310	2300 - 2310	2300 - 2310	(d), (p)
13 cm	2390 - 2450	2390 - 2450	2390 -	(d), (e), (p)

Wavelength band	ITU region 1	ITU region 2	ITU region 3	Sharing requirements see §97.303 (paragraph)
			2450	
SHF	GHz	GHz	GHz	
5 cm	5.650 - 5.850	5.650 - 5.925	5.650 - 5.850	(a), (b), (e), (r)
3 cm	10.0 - 10.5	10.0 - 10.5	10.0 - 10.5	(a), (b), (k)
1.2 cm	24.00 - 24.25	24.00 - 24.25	24.00 - 24.25	(b), (d), (e)
EHF	GHz	GHz	GHz	
6 mm	47.0 - 47.2	47.0 - 47.2	47.0 - 47.2	
4 mm	76 - 81	76 - 81	76 - 81	(c), (f), (s)
2.5 mm	122.25 - 123.00	122.25 - 123.00	122.25 - 123.00	(e), (t)
2 mm	134 - 141	134 - 141	134 - 141	(c), (f)
1 mm	241 - 250	241 - 250	241 - 250	(c), (e), (f)
	Above 275	Above 275	Above 275	(f)

(b) For a station having a control operator who has been granted an Amateur Extra Class operator license, who holds a CEPT radio amateur license, or who holds a Class 1 IARP license:

Wavelength band	ITU Region 1	ITU Region 2	ITU Region 3	Sharing requirements see §97.303 (paragraph)
LF	kHz	kHz	kHz	
2200 m	135.7 - 137.8	135.7 - 137.8	135.7 - 137.8	(a), (g).
160 m	1810 - 1850	1800 - 2000	1800 - 2000	(a)
630 m	472 - 479	472 - 479	472 - 479	(g).
HF	MHz	MHz	MHz	
80 m	3.500 - 3.600	3.500 - 3.600	3.500 - 3.600	(a)
75 m	3.600 - 3.800	3.600 - 4.000	3.600 - 3.900	(a)
60 m		See §97.303(h)		(h)
40 m	7.000 - 7.200	7.000 - 7.300	7.000 - 7.200	(i)
30 m	10.100 - 10.150	10.100 - 10.150	10.100 - 10.150	(j)
20 m	14.000 - 14.350	14.000 - 14.350	14.000 - 14.350	
Wavelength band	ITU Region 1	ITU Region 2	ITU Region 3	Sharing requirements see §97.303

Wavelength band	ITU Region 1	ITU Region 2	ITU Region 3	Sharing requirements see §97.303 (paragraph)
				(paragraph)
17 m	18.068 - 18.168	18.068 - 18.168	18.068 - 18.168	
15 m	21.000 - 21.450	21.000 - 21.450	21.000 - 21.450	
12 m	24.890 - 24.990	24.890 - 24.990	24.890 - 24.990	
10 m	28.000 - 29.700	28.000 - 29.700	28.000 - 29.700	

(c) For a station having a control operator who has been granted an operator license of Advanced Class:

Wavelength band	ITU Region 1	ITU Region 2	ITU Region 3	Sharing requirements see §97.303 (paragraph)
LF	kHz	kHz	kHz	
2200 m	135.7 - 137.8	135.7 - 137.8	135.7 - 137.8	(a), (g).
MF	kHz	kHz	kHz	
160 m	1810 - 1850	1800 - 2000	1800 - 2000	(a)
630 m	472 - 479	472 - 479	472 - 479	(g).

Wavelength band	ITU Region 1	ITU Region 2	ITU Region 3	Sharing requirements see §97.303 (paragraph)
HF	MHz	MHz	MHz	
80 m	3.525 - 3.600	3.525 - 3.600	3.525 - 3.600	(a)
75 m	3.700 - 3.800	3.700 - 4.000	3.700 - 3.900	(a)
60 m		See §97.303(h)		(h)
40 m	7.025 - 7.200	7.025 - 7.300	7.025 - 7.200	(i)
30 m	10.100 - 10.150	10.100 - 10.150	10.100 - 10.150	(j)
20 m	14.025 - 14.150	14.025 - 14.150	14.025 - 14.150	
20 m	14.175 - 14.350	14.175 - 14.350	14.175 - 14.350	
17 m	18.068 - 18.168	18.068 - 18.168	18.068 - 18.168	
15 m	21.025 - 21.200	21.025 - 21.200	21.025 - 21.200	
15 m	21.225 - 21.450	21.225 - 21.450	21.225 - 21.450	
12 m	24.890 - 24.990	24.890 - 24.990	24.890 - 24.990	
10 m	28.000 - 29.700	28.000 - 29.700	28.000 - 29.700	

(d) For a station having a control operator who has been granted an operator license of General Class:

Wavelength band	ITU Region 1	ITU Region 2	ITU Region 3	Sharing requirements see §97.303 (paragraph)
LF	kHz	kHz	kHz	
2200 m	135.7 - 137.8	135.7 - 137.8	135.7 - 137.8	(a), (g).
MF	kHz	kHz	kHz	Sharing requirements see § 97.303 (paragraph)
160 m	1810 - 1850	1800 - 2000	1800 - 2000	(a)
630 m	472 - 479	472 - 479	472 - 479	(g).
HF	MHz	MHz	MHz	
80 m	3.525 - 3.600	3.525 - 3.600	3.525 - 3.600	(a)
75 m		3.800 - 4.000	3.800 - 3.900	(a)
60 m	See §97.303(h)			(h)

Wavelength band	ITU Region 1	ITU Region 2	ITU Region 3	Sharing requirements see §97.303 (paragraph)
40 m	7.025 - 7.125	7.025 - 7.125	7.025 - 7.125	(i)
	7.175 - 7.200	7.175 - 7.300	7.175 - 7.200	(i)
30 m	10.100 - 10.150	10.100 - 10.150	10.100 - 10.150	(j)
20 m	14.025 - 14.150	14.025 - 14.150	14.025 - 14.150	
	14.225 - 14.350	14.225 - 14.350	14.225 - 14.350	
17 m	18.068 - 18.168	18.068 - 18.168	18.068 - 18.168	
15 m	21.025 - 21.200	21.025 - 21.200	21.025 - 21.200	
	21.275 - 21.450	21.275 - 21.450	21.275 - 21.450	
12 m	24.890 - 24.990	24.890 - 24.990	24.890 - 24.990	
10 m	28.000 - 29.700	28.000 - 29.700	28.000 - 29.700	

(e) For a station having a control operator who has been granted an operator license of Novice Class or Technician Class:

Wavelength band	ITU region 1	ITU region 2	ITU region 3	Sharing requirements see §97.303 (paragraph)
HF	MHz	MHz	MHz	

Wavelength band	ITU region 1	ITU region 2	ITU region 3	Sharing requirements see §97.303 (paragraph)
80 m	3.525 - 3.600	3.525 - 3.600	3.525 - 3.600	(a)
40 m	7.025 - 7.125	7.025 - 7.125	7.025 - 7.125	(i)
15 m	21.025 - 21.200	21.025 - 21.200	21.025 - 21.200	
10 m	28.0 - 28.5	28.0 - 28.5	28.0 - 28.5	
VHF	MHz	MHz	MHz	
1.25 m		222 - 225		(a)
UHF	MHz	MHz	MHz	Sharing requirements see §97.303 (paragraph)
23 cm	1270 - 1295	1270 - 1295	1270 - 1295	(d), (o)

[75 FR 27201, May 14, 2010, as amended at 75 FR 78171, Dec. 15, 2010; 80 FR 38911, July 7, 2015; 82 FR 27214, June 14, 2017]

§97.303 Frequency sharing requirements.

The following paragraphs summarize the frequency sharing requirements that apply to amateur stations transmitting in the frequency bands specified in § 97.301 of this part. Each frequency band allocated to the amateur service is designated as either a secondary service or a primary service. A station in a secondary service must not cause harmful interference to, and must accept interference from, stations in a primary service.

(a) Where, in adjacent ITU Regions or sub-Regions, a band of frequencies is allocated to different services of the same category (*i.e.*, primary or secondary services), the basic principle is the equality of right to operate. Accordingly, stations of each service in one Region or sub-Region must operate so as not to cause harmful interference to any service of the same or higher category in the other Regions or sub-Regions.

(b) Amateur stations transmitting in the 70 cm band, the 33 cm band, the 23 cm band, the 5 cm band, the 3 cm band, or the 24.05-24.25 GHz segment must not cause harmful interference to, and must accept interference from, stations authorized by the United States Government in the radiolocation service.

(c) Amateur stations transmitting in the 76-81 GHz segment, the 136-141 GHz segment, or the 241-248 GHz segment must not cause harmful interference to, and must accept interference from, stations authorized by the United States Government, the FCC, or other nations in the radiolocation service.

(d) Amateur stations transmitting in the 430-450 MHz segment, the 23 cm band, the 3.3-3.4 GHz segment, the 5.65-5.85 GHz segment, the 13 cm band, or the 24.05-24.25 GHz segment, must not cause harmful interference to, and must accept interference from, stations authorized by other nations in the radiolocation service.

(e) Amateur stations receiving in the 33 cm band, the 2400-2450 MHz segment, the 5.725-5.875 GHz segment, the 1.2 cm band, the 2.5 mm band, or the 244-246 GHz segment must accept interference from industrial, scientific, and medical (ISM) equipment.

(f) Amateur stations transmitting in the following segments must not cause harmful interference to radio astronomy stations: 76-81 GHz, 136-141 GHz, 241-248 GHz, 275-323 GHz, 327-371 GHz, 388-424 GHz, 426-442 GHz, 453-510 GHz, 623-711 GHz, 795-909 GHz, or 926-945 GHz. In addition, amateur stations transmitting in the following segments must not cause harmful interference to stations in the Earth exploration-satellite service (passive) or the space research service (passive): 275-286 GHz, 296-306 GHz, 313-356 GHz, 361-365 GHz, 369-392 GHz, 397-399 GHz, 409-411 GHz, 416-434 GHz, 439-467 GHz, 477-502 GHz, 523-527 GHz, 538-581 GHz, 611-630 GHz, 634-654 GHz, 657-692 GHz, 713-718 GHz, 729-733 GHz, 750-754 GHz, 771-776 GHz, 823-846 GHz, 850-854 GHz, 857-862 GHz, 866-882 GHz, 905-928 GHz, 951-956 GHz, 968-973 GHz and 985-990 GHz.

(g) In the 2200 m and 630 m bands:

(1) Amateur stations in the 135.7-137.8 kHz (2200 m) and 472-479 kHz (630 m) bands shall only operate at fixed locations. Amateur stations shall not operate within a horizontal distance of one kilometer from a transmission line that conducts a power line carrier (PLC) signal in the 135.7-137.8 kHz or 472-479 kHz bands. Horizontal distance is measured from the station's antenna to the closest point on the transmission line.

(2) Prior to commencement of operations in the 135.7-137.8 kHz (2200 m) and/or 472-479 kHz (630 m) bands, amateur operators shall notify the Utilities Telecom Council (UTC) of their intent to operate by submitting their call signs, intended band or bands of

operation, and the coordinates of their antenna's fixed location. Amateur stations will be permitted to commence operations after the 30-day period unless UTC notifies the station that its fixed location is located within one kilometer of PLC systems operating in the same or overlapping frequencies.

(3) Amateur stations in the 135.7-137.8 kHz (2200 m) band shall not cause harmful interference to, and shall accept interference from:

(i) Stations authorized by the United States Government in the fixed and maritime mobile services;

(ii) Stations authorized by other nations in the fixed, maritime mobile, and radionavigation service.

(4) Amateur stations in the 472-479 kHz (630 m) band shall not cause harmful interference to, and shall accept interference from:

(i) Stations authorized by the FCC in the maritime mobile service;

(ii) Stations authorized by other nations in the maritime mobile and aeronautical radionavigation services.

(5) Amateur stations causing harmful interference shall take all necessary measures to eliminate such interference - including temporary or permanent termination of transmissions.

(h) *60 m band:*

(1) In the 5330.5-5406.4 kHz band (60 m band), amateur stations may transmit only on the five center frequencies specified in the table below. In order to meet this requirement, control operators of stations transmitting phone, data, and RTTY emissions (emission designators 2K80J3E, 2K80J2D, and 60H0J2B, respectively) may set the carrier frequency 1.5 kHz below the center frequency as specified in the table below. For CW emissions (emission designator 150HA1A), the carrier frequency is set to the center frequency. Amateur operators shall ensure that their emissions do not occupy more than 2.8 kHz centered on each of these center frequencies.

60 M BAND FREQUENCIES (KHZ)

Carrier	Center
5330.5	5332.0
5346.5	5348.0
5357.0	5358.5
5371.5	5373.0
5403.5	5405.0

Amateur stations transmitting on the 60 m band must not cause harmful interference to, and must accept interference from, stations authorized by:

(i) The United States (NTIA and FCC) and other nations in the fixed service; and

(ii) Other nations in the mobile except aeronautical mobile service.

(i) Amateur stations transmitting in the 7.2-7.3 MHz segment must not cause harmful interference to, and must accept interference from, international broadcast stations whose programming is intended for use within Region 1 or Region 3.

(j) Amateur stations transmitting in the 30 m band must not cause harmful interference to, and must accept interference from, stations by other nations in the fixed service. The licensee of the amateur station must make all necessary adjustments, including termination of transmissions, if harmful interference is caused.

(k) For amateur stations located in ITU Regions 1 and 3: Amateur stations transmitting in the 146-148 MHz segment or the 10.00-10.45 GHz segment must not cause harmful interference to, and must accept interference from, stations of other nations in the fixed and mobile services.

(l) *In the 219-220 MHz segment:*

(1) Use is restricted to amateur stations participating as forwarding stations in fixed point-to-point digital message forwarding systems, including intercity packet backbone networks. It is not available for other purposes.

(2) Amateur stations must not cause harmful interference to, and must accept interference from, stations authorized by:

(i) The FCC in the Automated Maritime Telecommunications System (AMTS), the 218-219 MHz Service, and the 220 MHz Service, and television stations broadcasting on channels 11 and 13; and

(ii) Other nations in the fixed and maritime mobile services.

(3) No amateur station may transmit unless the licensee has given written notification of the station's specific geographic location for such transmissions in order to be incorporated into a database that has been made available to the public. The notification must be given at least 30 days prior to making such transmissions. The notification must be given to: The American Radio Relay League, Inc., 225 Main Street, Newington, CT 06111-1494.

(4) No amateur station may transmit from a location that is within 640 km of an AMTS coast station that operates in the 217-218 MHz and 219-220 MHz bands unless the amateur station licensee has given written notification of the station's specific geographic location for such transmissions to the AMTS licensee. The notification must be given at least 30 days prior to making such transmissions. The location of AMTS coast stations using the 217-218/219-220 MHz channels may be obtained as noted in paragraph (l)(3) of this section.

(5) No amateur station may transmit from a location that is within 80 km of an AMTS coast station that uses frequencies in the 217-218 MHz and 219-220 MHz bands unless that amateur station licensee holds written approval from that AMTS licensee. The location of AMTS coast stations using the 217-218/219-220 MHz channels may be obtained as noted in paragraph (l)(3) of this section.

(m) *In the 70 cm band:*

(1) No amateur station shall transmit from north of Line A in the 420-430 MHz segment. See § 97.3(a) for the definition of Line A.

(2) Amateur stations transmitting in the 420-430 MHz segment must not cause harmful interference to, and must accept interference from, stations authorized by the FCC in the land mobile service within 80.5 km of Buffalo, Cleveland, and Detroit. See § 2.106, footnote US230 for specific frequencies and coordinates.

(3) Amateur stations transmitting in the 420-430 MHz segment or the 440-450 MHz segment must not cause harmful interference to, and must accept interference from, stations authorized by other nations in the fixed and mobile except aeronautical mobile services.

(n) *In the 33 cm band:*

(1) Amateur stations must not cause harmful interference to, and must accept interference from, stations authorized by:

(i) The United States Government;

(ii) The FCC in the Location and Monitoring Service; and

(iii) Other nations in the fixed service.

(2) No amateur station shall transmit from those portions of Texas and New Mexico that are bounded by latitudes 31°41' and 34°30' North and longitudes 104°11' and 107°30' West; or from outside of the United States and its Region 2 insular areas.

(3) No amateur station shall transmit from those portions of Colorado and Wyoming that are bounded by latitudes 39° and 42° North and longitudes 103° and 108° West in the following segments: 902.4-902.6 MHz, 904.3-904.7 MHz, 925.3-925.7 MHz, and 927.3-927.7 MHz.

(o) Amateur stations transmitting in the 23 cm band must not cause harmful interference to, and must accept interference from, stations authorized by:

(1) The United States Government in the aeronautical radionavigation, Earth exploration-satellite (active), or space research (active) services;

(2) The FCC in the aeronautical radionavigation service; and

(3) Other nations in the Earth exploration-satellite (active), radionavigation-satellite (space-to-Earth) (space-to-space), or space research (active) services.

(p) *In the 13 cm band:*

(1) Amateur stations must not cause harmful interference to, and must accept interference from, stations authorized by other nations in fixed and mobile services.

(2) Amateur stations transmitting in the 2305-2310 MHz segment must not cause harmful interference to, and must accept interference from, stations authorized by the FCC in the fixed, mobile except aeronautical mobile, and radiolocation services.

(q) [Reserved]

(r) *In the 5 cm band:*

(1) Amateur stations transmitting in the 5.650-5.725 GHz segment must not cause harmful interference to, and must accept interference from, stations authorized by other nations in the mobile except aeronautical mobile service.

(2) Amateur stations transmitting in the 5.850-5.925 GHz segment must not cause harmful interference to, and must accept interference from, stations authorized by the FCC and other nations in the fixed-satellite (Earth-to-space) and mobile services and also stations authorized by other nations in the fixed service. In the United States, the use of mobile service is restricted to Dedicated Short Range Communications operating in the Intelligent Transportation System.

(s) [Reserved]

(t) Amateur stations transmitting in the 2.5 mm band must not cause harmful interference to, and must accept interference from, stations authorized by the United States Government, the FCC, or other nations in the fixed, inter-satellite, or mobile services.

Note to § 97.303:

The Table of Frequency Allocations contains the complete, unabridged, and legally binding frequency sharing requirements that pertain to the Amateur Radio Service. *See* 47 CFR 2.104, 2.105, and 2.106. The United States, Puerto Rico, and the U.S. Virgin Islands are in Region 2 and other U.S. insular areas are in either Region 2 or 3; see appendix 1 to part 97.

[75 FR 27203, May 14, 2010, as amended at 77 FR 5412, Feb. 3, 2012; 80 FR 38912, July 7, 2015; 82 FR 27215, June 14, 2017; 82 FR 43872, Sept. 20, 2017; 85 FR 64068, Oct. 9, 2020]

§97.305 Authorized emission types.

(a) Except as specified elsewhere in this part, an amateur station may transmit a CW emission on any frequency authorized to the control operator.

(b) A station may transmit a test emission on any frequency authorized to the control operator for brief periods for experimental purposes, except that no pulse modulation emission may be transmitted on any frequency where pulse is not specifically authorized and no SS modulation emission may be transmitted on any frequency where SS is not specifically authorized.

(c) A station may transmit the following emission types on the frequencies indicated, as authorized to the control operator, subject to the standards specified in § 97.307(f):

Wavelength band	Frequencies	Emission types authorized	Standards see §97.307(f), paragraph:
LF:			
2200 m	Entire band	RTTY, data	(f)(3).
2200 m	Entire band	Phone, image	(f)(1), (2).
MF:			
630 m	Entire band	RTTY, data	(f) (3).
630 m	Entire band	Phone, image	(f) (1), (2).
160 m	Entire band	RTTY, data	(f) (3).
160 m	Entire band	Phone, image	(f) (1), (2).
HF:			
80 m	Entire band	RTTY, data	(f) (3), (9).
75 m	Entire band	Phone, image	(f) (1), (2).
60 m	5.332, 5.348, 5.3585, 5.373 and 5.405 MHz	Phone, RTTY, data	(f) (14).
40 m	7.000 - 7.100 MHz	RTTY, data	(f) (3), (9)
40 m	7.075 - 7.100 MHz	Phone, image	(f) (1), (2), (9), (11)
40 m	7.100 - 7.125 MHz	RTTY, data	(f) (3), (9)
40 m	7.125 - 7.300 MHz	Phone, image	(f) (1), (2)

Wavelength band	Frequencies	Emission types authorized	Standards see §97.307(f), paragraph:
30 m	Entire band	RTTY, data	(f) (3).
20 m	14.00 - 14.15 MHz	RTTY, data	(f) (3).
	14.15 - 14.35 MHz	Phone, image	(f) (1), (2).
17 m	18.068 - 18.110 MHz	RTTY, data	(f) (3).
	18.110 - 18.168 MHz	Phone, image	(f) (1), (2).
15 m	21.0 - 21.2 MHz	RTTY, data	(f) (3), (9).
	21.20 - 21.45 MHz	Phone, image	(f) (1), (2).
12 m	24.89 - 24.93 MHz	RTTY, data	(f) (3).
	24.93 - 24.99 MHz	Phone, image	(f) (1), (2).
10 m	28.0 - 28.3 MHz	RTTY, data	(f) (4).
	28.3 - 28.5 MHz	Phone, image	(f) (1), (2), (10).
	28.5 - 29.0 MHz	Phone, image	(f) (1), (2).
	29.0 - 29.7 MHz	Phone, image	(f) (2).
VHF:			
6 m	50.1 - 51.0 MHz	MCW, phone, image, RTTY, data	(f) (2), (5).
	51.0 - 54.0 MHz	MCW, phone, image, RTTY, data, test	(f) (2), (5), (8).
2 m	144.1 - 148.0 MHz	MCW, phone, image, RTTY, data, test	(f) (2), (5), (8).
1.25 m	219 - 220 MHz	Data	(f) (13)
	222 - 225 MHz	RTTY, data, test MCW, phone, SS, image	(f) (2), (6), (8)
UHF:			
70 cm	Entire band	MCW, phone, image, RTTY, data, SS, test	(f) (6), (8).

Wavelength band	Frequencies	Emission types authorized	Standards see §97.307(f), paragraph:
33 cm	Entire band	MCW, phone, image, RTTY, data, SS, test, pulse	(f) (7), (8), and (12).
23 cm	Entire band	MCW, phone, image, RTTY, data, SS, test	(f) (7), (8), and (12).
13 cm	Entire band	MCW, phone, image, RTTY, data, SS, test, pulse	(f) (7), (8), and (12).
SHF:			
9 cm	Entire band	MCW, phone, image, RTTY, data, SS, test, pulse	(f) (7), (8), and (12).
5 cm	Entire band	MCW, phone, image, RTTY, data, SS, test, pulse	(f) (7), (8), and (12).
3 cm	Entire band	MCW, phone, image, RTTY, data, SS, test	(f) (7), (8), and (12).
1.2 cm	Entire band	MCW, phone, image, RTTY, data, SS, test, pulse	(f) (7), (8), and (12).
EHF:			
6 mm	Entire band	MCW, phone, image, RTTY, data, SS, test, pulse	(f) (7), (8), and (12).
4 mm	Entire band	MCW, phone, image, RTTY, data, SS, test, pulse	(f) (7), (8), and (12).
2.5 mm	Entire band	MCW, phone, image, RTTY, data, SS, test, pulse	(f) (7), (8), and (12).
2 mm	Entire band	MCW, phone, image, RTTY, data, SS, test, pulse	(f) (7), (8), and (12).
1mm	Entire band	MCW, phone, image, RTTY, data, SS, test, pulse	(f) (7), (8), and (12).
< 1mm	Above 275 GHz	MCW, phone, image, RTTY, data, SS, test, pulse	(f) (7), (8), and (12).

[54 FR 25857, June 20, 1989; 54 FR 39536, Sept. 27, 1989; 55 FR 22013, May 30, 1990, as amended at 55 FR 30457, July 26, 1990; 60 FR 15688, Mar. 27, 1995; 64 FR 51471, Sept. 23, 1999; 71 FR 66465, Nov. 15, 2006; 75 FR 27204, May 14, 2010; 77 FR 5412, Feb. 3, 2012; 82 FR 27215, June 14, 2017; 85 FR 64069, Oct. 9, 2020; 88 FR 85127, Dec. 7, 2023; 88 FR 85128, Dec. 7, 2023]]

§97.307 Emission standards.

(a) No amateur station transmission shall occupy more bandwidth than necessary for the information rate and emission type being transmitted, in accordance with good amateur practice.

(b) Emissions resulting from modulation must be confined to the band or segment available to the control operator. Emissions outside the necessary bandwidth must not cause splatter or keyclick interference to operations on adjacent frequencies.

(c) All spurious emissions from a station transmitter must be reduced to the greatest extent practicable. If any spurious emission, including chassis or power line radiation, causes harmful interference to the reception of another radio station, the licensee of the interfering amateur station is required to take steps to eliminate the interference, in accordance with good engineering practice.

(d) For transmitters installed after January 1, 2003, the mean power of any spurious emission from a station transmitter or external RF power amplifier transmitting on a frequency below 30 MHz must be at least 43 dB below the mean power of the fundamental emission. For transmitters installed on or before January 1, 2003, the mean power of any spurious emission from a station transmitter or external RF power amplifier transmitting on a frequency below 30 MHz must not exceed 50 mW and must be at least 40 dB below the mean power of the fundamental emission. For a transmitter of mean power less than 5 W installed on or before January 1, 2003, the attenuation must be at least 30 dB. A transmitter built before April 15, 1977, or first marketed before January 1, 1978, is exempt from this requirement.

(e) The mean power of any spurious emission from a station transmitter or external RF power amplifier transmitting on a frequency between 30-225 MHz must be at least 60 dB below the mean power of the fundamental. For a transmitter having a mean power of 25 W or less, the mean power of any spurious emission supplied to the antenna transmission line must not exceed 25 μW and must be at least 40 dB below the mean power of the fundamental emission, but need not be reduced below the power of 10 μW. A transmitter built before April 15, 1977, or first marketed before January 1, 1978, is exempt from this requirement.

(f) The following standards and limitations apply to transmissions on the frequencies specified in § 97.305(c).

(1) No angle-modulated emission may have a modulation index greater than 1 at the highest modulation frequency.

(2) No non-phone emission shall exceed the bandwidth of a communications quality phone emission of the same modulation type. The total bandwidth of an independent sideband emission (having B as the first symbol), or a multiplexed image and phone emission, shall not exceed that of a communications quality A3E emission.

(3) Only a RTTY or data emission using a specified digital code listed in § 97.309(a) may be transmitted. The authorized bandwidth is 2.8 kHz except in the 2200 m band and 630 m band. In the 2200 m band and the 630 m band the symbol rate must not exceed 300 bauds, or for frequency-shift keying, the frequency shift between mark and space must not exceed 1 kHz.

(4) [Reserved]

(5) A RTTY, data or multiplexed emission using a specified digital code listed in § 97.309(a) may be transmitted. The symbol rate must not exceed 19.6 kilobauds. A RTTY, data or multiplexed emission using an unspecified digital code under the limitations listed in § 97.309(b) also may be transmitted. The authorized bandwidth is 20 kHz.

(6) A RTTY, data or multiplexed emission using a specified digital code listed in § 97.309(a) may be transmitted. The symbol rate must not exceed 56 kilobauds. A RTTY, data or multiplexed emission using an unspecified digital code under the limitations listed in § 97.309(b) also may be transmitted. The authorized bandwidth is 100 kHz.

(7) A RTTY, data or multiplexed emission using a specified digital code listed in § 97.309(a) or an unspecified digital code under the limitations listed in § 97.309(b) may be transmitted.

(8) A RTTY or data emission having designators with A, B, C, D, E, F, G, H, J or R as the first symbol; 1, 2, 7, 9 or X as the second symbol; and D or W as the third symbol is also authorized.

(9) A station having a control operator holding a Novice or Technician Class operator license may only transmit a CW emission using the international Morse code.

(10) A station having a control operator holding a Novice Class operator license or a Technician Class operator license may only transmit a CW emission using the international Morse code or phone emissions J3E and R3E.

(11) Phone and image emissions may be transmitted only by stations located in ITU Regions 1 and 3, and by stations located within ITU Region 2 that are west of 130° West longitude or south of 20° North latitude.

(12) Emission F8E may be transmitted.

(13) A data emission using an unspecified digital code under the limitations listed in § 97.309(b) also may be transmitted. The authorized bandwidth is 100 kHz.

(14) *In the 60 m band:*

(i) A station may transmit only phone, RTTY, data, and CW emissions using the emission designators and any additional restrictions that are specified in the table below (except that the use of a narrower necessary bandwidth is permitted):

60 M Band Emission Requirements

Emission type	Emission designator	Restricted to:
Phone	2K80 J3E	Upper sideband transmissions (USB).
Data	2K80 J2D	USB (for example, PACTOR-III).
RTTY	60H0 J2B	USB (for example, PSK31).
CW	150H A1A	Morse telegraphy by means of on-off keying.

(ii) The following requirements also apply:

(A) When transmitting the phone, RTTY, and data emissions, the suppressed carrier frequency may be set as specified in §97.303(h).

(B) The control operator of a station transmitting data or RTTY emissions must exercise care to limit the length of transmission so as to avoid causing harmful interference to United States Government stations.

[54 FR 25857, June 20, 1989; 54 FR 30823, July 24, 1989, as amended at 54 FR 39537, Sept. 27, 1989; 60 FR 15688, Mar. 27, 1995; 65 FR 6550, Feb. 10, 2000; 69 FR 24997, May 5, 2004; 77 FR 5412, Feb. 3, 2012; 79 FR 35291, June 20, 2014; 88 FR 85128, Dec. 7, 2023]

§97.309 RTTY and data emission codes.

(a) Where authorized by §§97.305(c) and 97.307(f) of the part, an amateur station may transmit a RTTY or data emission using the following specified digital codes:

(1) The 5-unit, start-stop, International Telegraph Alphabet No. 2, code defined in ITU-T Recommendation F.1, Division C (commonly known as "Baudot").

(2) The 7-unit code specified in ITU-R Recommendations M.476-5 and M.625-3 (commonly known as "AMTOR").

(3) The 7-unit, International Alphabet No. 5, code defined in IT-T Recommendation T.50 (commonly known as "ASCII").

(4) An amateur station transmitting a RTTY or data emission using a digital code specified in this paragraph may use any technique whose technical characteristics have been documented publicly, such as CLOVER, G-TOR, or PacTOR, for the purpose of facilitating communications.

(b) Where authorized by §§97.305(c) and 97.307(f), a station may transmit a RTTY or data emission using an unspecified digital code, except to a station in a country with which the United States does not have an agreement permitting the code to be used. RTTY and data emissions using unspecified digital codes must not be transmitted for the purpose of obscuring the meaning of any communication. When deemed necessary by a Regional Director to assure compliance with the FCC Rules, a station must:

(1) Cease the transmission using the unspecified digital code;

(2) Restrict transmissions of any digital code to the extent instructed;

(3) Maintain a record, convertible to the original information, of all digital communications transmitted.

[54 FR 25857, June 20, 1989, as amended at 54 FR 39537, Sept. 27, 1989; 56 FR 56172, Nov. 1, 1991; 60 FR 55486, Nov. 1, 1995; 71 FR 25982, May 3, 2006; 71 FR 66465, Nov. 15, 2006; 80 FR 53753, Sept. 8, 2015]

§97.311 SS emission types.

(a) SS emission transmissions by an amateur station are authorized only for communications between points within areas where the amateur service is regulated by the FCC and between an area where the amateur service is regulated by the FCC and an amateur station in another country that permits such communications. SS emission transmissions must not be used for the purpose of obscuring the meaning of any communication.

(b) A station transmitting SS emissions must not cause harmful interference to stations employing other authorized emissions, and must accept all interference caused by stations employing other authorized emissions.

(c) When deemed necessary by a Regional Director to assure compliance with this part, a station licensee must:

(1) Cease SS emission transmissions;

(2) Restrict SS emission transmissions to the extent instructed; and

(3) Maintain a record, convertible to the original information (voice, text, image, etc.) of all spread spectrum communications transmitted.

[64 FR 51471, Sept. 23, 1999, as amended at 76 FR 17569, Mar. 30, 2011; 80 FR 53753, Sept. 8, 2015]

§97.313 Transmitter power standards.

(a) An amateur station must use the minimum transmitter power necessary to carry out the desired communications.

(b) No station may transmit with a transmitter power exceeding 1.5 kW PEP.

(c) No station may transmit with a transmitter power output exceeding 200 W PEP:

(1) On the 10.10 - 10.15 MHz segment;

(2) On the 3.525 - 3.60 MHz, 7.025 - 7.125 MHz, 21.025 - 21.20 MHz, and 28.0 - 28.5 MHz segment when the control operator is a Novice Class operator or a Technician Class operator; or

(3) The 7.050 - 7.075 MHz segment when the station is within ITU Regions 1 or 3.

(d) No station may transmit with a transmitter power exceeding 25 W PEP on the VHF 1.25 m band when the control operator is a Novice operator.

(e) No station may transmit with a transmitter power exceeding 5 W PEP on the UHF 23 cm band when the control operator is a Novice operator.

(f) No station may transmit with a transmitter power exceeding 50 W PEP on the UHF 70 cm band from an area specified in paragraph (a) of footnote US270 in §2.106, unless expressly authorized by the FCC after mutual agreement, on a case-by-case basis, between the Regional Director of the applicable field facility and the military area frequency coordinator at the applicable military base. [*Editor's note: US270 from §2.106 is reproduced at the end of Part 97.*] An Earth station or telecommand station, however, may transmit on the 435 - 438 MHz segment with a maximum of 611 W effective radiated power (1 kW equivalent isotropically radiated power) without the authorization otherwise required. The transmitting antenna elevation angle between the lower half-power (−3 dB relative to the peak or antenna bore sight) point and the horizon must always be greater than 10°.

(g) No station may transmit with a transmitter power exceeding 50 W PEP on the 33 cm band from within 241 km of the boundaries of the White Sands Missile Range. Its boundaries are those portions of Texas and New Mexico bounded on the south by latitude 31°41′ North, on the east by longitude 104°11′ West, on the north by latitude 34°30′ North, and on the west by longitude 107°30′ West.

(h) No station may transmit with a transmitter power exceeding 50 W PEP on the 219 - 220 MHz segment of the 1.25 m band.

(i) No station may transmit with an effective radiated power (ERP) exceeding 100 W PEP on the 60 m band. For the purpose of computing ERP, the transmitter PEP will be multiplied by the antenna gain relative to a half-wave dipole antenna. A half-wave dipole antenna will be presumed to have a gain of 1 (0 dBd). Licensees using other antennas must maintain in their station records either the antenna manufacturer's data on the antenna gain or calculations of the antenna gain.

(j) No station may transmit with a transmitter output exceeding 10 W PEP when the station is transmitting a SS emission type.

(k) No station may transmit in the 135.7 - 137.8 kHz (2200 m) band with a transmitter power exceeding 1.5 kW PEP or a radiated power exceeding 1 W EIRP.

(l) No station may transmit in the 472 - 479 kHz (630 m) band with a transmitter power exceeding 500 W PEP or a radiated power exceeding 5 W EIRP, except that in Alaska, stations located within 800 kilometers of the Russian Federation may not transmit with a radiated power exceeding 1 W EIRP.

(m) No station may transmit with a peak equivalent isotropically radiated power (EIRP) exceeding 316 W in the 76 - 81 GHz (4 mm) band.

[54 FR 25857, June 20, 1989, as amended at 56 FR 37161, Aug. 5, 1991; 56 FR 3043, Jan. 28, 1991; 60 FR 15688, Mar. 27, 1995; 65 FR 6550, Feb. 10, 2000; 71 FR 66465, Nov. 15, 2006; 75 FR 27204, May 14, 2010; 75 FR 78171, Dec. 15, 2010; 76 FR 17569, Mar. 30, 2011; 77 FR 5413, Feb. 3, 2012; 80 FR 53753, Sept. 8, 2015; 82 FR 27216, June 14, 2017; 82 FR 43872, Sept. 20, 2017]

§97.315 Certification of external RF power amplifiers.

(a) Any external RF power amplifier (see §2.815 of the FCC Rules) manufactured or imported for use at an amateur radio station must be certificated for use in the amateur service in accordance with subpart J of part 2 of the FCC Rules. No amplifier capable of operation below 144 MHz may be constructed or modified by a non-amateur service licensee without a grant of certification from the FCC.

(b) The requirement of paragraph (a) does not apply if one or more of the following conditions are met:

(1) The amplifier is constructed or modified by an amateur radio operator for use at an amateur station.

(2) The amplifier was manufactured before April 28, 1978, and has been issued a marketing waiver by the FCC, or the amplifier was purchased before April 28, 1978, by an amateur radio operator for use at that operator's station.

(3) The amplifier is sold to an amateur radio operator or to a dealer, the amplifier is purchased in used condition by a dealer, or the amplifier is sold to an amateur radio operator for use at that operator's station.

(c) Any external RF power amplifier appearing in the Commission's database as certificated for use in the amateur service may be marketed for use in the amateur service.

[71 FR 66465, Nov. 15, 2006]

§97.317 Standards for certification of external RF power amplifiers.

(a) To receive a grant of certification, the amplifier must:

(1) Satisfy the spurious emission standards of §97.307 (d) or (e) of this part, as applicable, when the amplifier is operated at the lesser of 1.5 kW PEP or its full output power and when the amplifier is placed in the "standby" or "off" positions while connected to the transmitter.

(2) Not be capable of amplifying the input RF power (driving signal) by more than 15 dB gain. Gain is defined as the ratio of the input RF power to the output RF power of the amplifier where both power measurements are expressed in peak envelope power or mean power.

(3) Exhibit no amplification (0 dB gain) between 26 MHz and 28 MHz.

(b) Certification shall be denied when:

(1) The Commission determines the amplifier can be used in services other than the Amateur Radio Service, or

(2) The amplifier can be easily modified to operate on frequencies between 26 MHz and 28 MHz.

[71 FR 66465, Nov. 15, 2006]

Subpart E—Providing Emergency Communications

§97.401   Operation during a disaster.

A station in, or within 92.6 km (50 nautical miles) of, Alaska may transmit emissions J3E [SSB phone with suppressed carrier] and R3E [SSB phone with reduced carrier] on the channel at 5.1675 MHz (assigned frequency 5.1689 MHz) for emergency communications. The channel must be shared with stations licensed in the Alaska-Private Fixed Service. The transmitter power must not exceed 150 W PEP. A station in, or within 92.6 km of, Alaska may transmit communications for tests and training drills necessary to ensure the establishment, operation, and maintenance of emergency communication systems.

[71 FR 66465, Nov. 15, 2006]

§97.403   Safety of life and protection of property.

No provision of these rules prevents the use by an amateur station of any means of radiocommunication at its disposal to provide essential communication needs in connection with the immediate safety of human life and immediate protection of property when normal communication systems are not available.

§97.405   Station in distress.

(a) No provision of these rules prevents the use by an amateur station in distress of any means at its disposal to attract attention, make known its condition and location, and obtain assistance.

(b) No provision of these rules prevents the use by a station, in the exceptional circumstances described in paragraph (a) of this section, of any means of radiocommunications at its disposal to assist a station in distress.

§97.407   Radio amateur civil emergency service.

(a) No station may transmit in RACES unless it is an FCC-licensed primary, club, or military recreation station and it is certified by a civil defense organization as registered with that organization. No person may be the control operator of an amateur station transmitting in RACES unless that person holds a FCC-issued amateur operator license and is certified by a civil defense organization as enrolled in that organization.

(b) The frequency bands and segments and emissions authorized to the control operator are available to stations transmitting communications in RACES on a shared basis with the amateur service. In the event of an emergency which necessitates invoking the President's War Emergency Powers under the provisions of section 706 of the Communications Act of 1934, as amended, 47 U.S.C. 606, amateur stations participating in RACES may only transmit on the frequency segments authorized pursuant to part 214 of this chapter.

(c) An amateur station registered with a civil defense organization may only communicate with the following stations upon authorization of the responsible civil defense official for the organization with which the amateur station is registered:

(1) An amateur station registered with the same or another civil defense organization; and

(2) A station in a service regulated by the FCC whenever such communication is authorized by the FCC.

(d) All communications transmitted in RACES must be specifically authorized by the civil defense organization for the area served. Only civil defense communications of the following types may be transmitted:

(1) Messages concerning impending or actual conditions jeopardizing the public safety, or affecting the national defense or security during periods of local, regional, or national civil emergencies;

(2) Messages directly concerning the immediate safety of life of individuals, the immediate protection of property, maintenance of law and order, alleviation of human suffering and need, and the combating of armed attack or sabotage;

(3) Messages directly concerning the accumulation and dissemination of public information or instructions to the civilian population essential to the activities of the civil defense organization or other authorized governmental or relief agencies; and

(4) Communications for RACES training drills and tests necessary to ensure the establishment and maintenance of orderly and efficient operation of the RACES as ordered by the responsible civil defense organization served. Such drills and tests may not exceed a total time of 1 hour per week. With the approval of the chief officer for emergency planning in the applicable State, Commonwealth, District or territory, however, such tests and drills may be conducted for a period not to exceed 72 hours no more than twice in any calendar year.

[75 FR 78171, Dec. 15, 2010]

Subpart F—Qualifying Examination Systems

§97.501 Qualifying for an amateur operator license.

Each applicant must pass an examination for a new amateur operator license grant and for each change in operator class. Each applicant for the class of operator license grant specified below must pass, or otherwise receive examination credit for, the following examination elements:

(a) Amateur Extra Class operator: Elements 2, 3, and 4;

(b) General Class operator: Elements 2 and 3;

(c) Technician Class operator: Element 2.

[65 FR 6550, Feb. 10, 2000, as amended at 72 FR 3082, Jan. 24, 2007]

§97.503 Element standards.

A written examination must be such as to prove that the examinee possesses the operational and technical qualifications required to perform properly the duties of an amateur service licensee. Each written examination must be comprised of a question set as follows:

(a) Element 2: 35 questions concerning the privileges of a Technician Class operator license. The minimum passing score is 26 questions answered correctly.

(b) Element 3: 35 questions concerning the privileges of a General Class operator license. The minimum passing score is 26 questions answered correctly.

(c) Element 4: 50 questions concerning the privileges of an Amateur Extra Class operator license. The minimum passing score is 37 questions answered correctly.

[54 FR 25857, June 20, 1989, as amended at 61 FR 41019, Aug. 7, 1996; 65 FR 6550, Feb. 10, 2000; 72 FR 3082, Jan. 24, 2007]

§97.505 Element credit.

(a) The administering VEs must give credit as specified below to an examinee holding any of the following license grants:

Operator class	Unexpired (or within the renewal grace period)	Expired and beyond the renewal grace period
(1) Amateur Extra	Not applicable	Elements 3 and 4.
(2) Advanced; General; or Technician granted before March 21, 1987	Elements 2 and 3	Element 3.
(3) Technician Plus; or Technician granted on or after March 21, 1987	Element 2	No credit.

(b) The administering VEs must give credit to an examinee holding a CSCE for each element the CSCE indicates the examinee passed within the previous 365 days.

[79 FR 35291, June 20, 2014]

§97.507 Preparing an examination.

(a) Each written question set administered to an examinee must be prepared by a VE holding an Amateur Extra Class operator license. A written question set may also be prepared for the following elements by a VE holding an operator license of the class indicated:

(1) Element 3: Advanced Class operator.

(2) Element 2: Advanced or General class operators.

(b) Each question set administered to an examinee must utilize questions taken from the applicable question pool.

(c) Each written question set administered to an examinee for an amateur operator license must be prepared, or obtained from a supplier, by the administering VEs according to instructions from the coordinating VEC.

[54 FR 25857, June 20, 1989, as amended at 58 FR 29126, May 19, 1993; 59 FR 54834, Nov. 2, 1994; 65 FR 6551, Feb. 10, 2000; 69 FR 24997, May 5, 2004; 79 FR 35291, June 20, 2014; 79 FR 52226, Sept. 3, 2014]

§97.509 Administering VE requirements.

(a) Each examination for an amateur operator license must be administered by a team of at least 3 VEs at an examination session coordinated by a VEC. The number of examinees at the session may be limited.

(b) Each administering VE must:

(1) Be accredited by the coordinating VEC;

(2) Be at least 18 years of age;

(3) Be a person who holds an amateur operator license of the class specified below:

(i) Amateur Extra, Advanced or General Class in order to administer a Technician Class operator license examination;

(ii) Amateur Extra or Advanced Class in order to administer a General Class operator license examination;

(iii) Amateur Extra Class in order to administer an Amateur Extra Class operator license examination.

(4) Not be a person whose grant of an amateur station license or amateur operator license has ever been revoked or suspended.

(c) Each administering VE must observe the examinee throughout the entire examination. The administering VEs are responsible for the proper conduct and necessary supervision of each examination. The administering VEs must immediately terminate the examination upon failure of the examinee to comply with their instructions.

(d) No VE may administer an examination to his or her spouse, children, grandchildren, stepchildren, parents, grandparents, stepparents, brothers, sisters, stepbrothers, stepsisters, aunts, uncles, nieces, nephews, and in-laws.

(e) No VE may administer or certify any examination by fraudulent means or for monetary or other consideration including reimbursement in any amount in excess of that permitted. Violation of this provision may result in the revocation of the grant of the VE's amateur station license and the suspension of the grant of the VE's amateur operator license.

(f) No examination that has been compromised shall be administered to any examinee. The same question set may not be re-administered to the same examinee.

(g) [Reserved]

(h) Upon completion of each examination element, the administering VEs must immediately grade the examinee's answers. For examinations administered remotely, the administering VEs must grade the examinee's answers at the earliest practical opportunity. The administering VEs are responsible for determining the correctness of the examinee's answers.

(i) When the examinee is credited for all examination elements required for the operator license sought, 3 VEs must certify that the examinee is qualified for the license grant and that the VEs have complied with these administering VE requirements. The certifying VEs are jointly and individually accountable for the proper administration of each examination element reported. The certifying VEs may delegate to other qualified VEs their authority, but not their accountability, to administer individual elements of an examination.

(j) When the examinee does not score a passing grade on an examination element, the administering VEs must return the application document to the examinee and inform the examinee of the grade.

(k) The administering VEs must accommodate an examinee whose physical disabilities require a special examination procedure. The administering VEs may require a physician's certification indicating the nature of the disability before determining which, if any, special procedures must be used.

(l) The administering VEs must issue a CSCE to an examinee who scores a passing grade on an examination element.

(m) After the administration of a successful examination for an amateur operator license, the administering VEs must submit the application document to the coordinating VEC according to the coordinating VEC's instructions.

[59 FR 54834, Nov. 2, 1994, as amended at 61 FR 9953, Mar. 12, 1996; 62 FR 17567, Apr. 10, 1997; 63 FR 68980, Dec. 14, 1998; 65 FR 6551, Feb. 10, 2000; 71 FR 66465, Nov. 15, 2006; 79 FR 35291. June 20, 2014]

§97.511 Examinee conduct.

Each examinee must comply with the instructions given by the administering VEs.

[59 FR 54835, Nov. 2, 1994]

§97.513 VE session manager requirements.

(a) A VE session manager may be selected by the VE team for each examination session. The VE session manager must be accredited as a VE by the same VEC that coordinates the examination session. The VE session manager may serve concurrently as an administering VE.

(b) The VE session manager may carry on liaison functions between the VE team and the coordinating VEC.

(c) The VE session manager may organize activities at an examination session.

[62 FR 17567, Apr. 10, 1997, as amended at 79 FR 35291, June 20, 2014]

§§97.515 - 97.517 [Reserved]

§97.519 Coordinating examination sessions.

(a) A VEC must coordinate the efforts of VEs in preparing and administering examinations.

(b) At the completion of each examination session, the coordinating VEC must collect applicant information and test results from the administering VEs. The coordinating VEC must:

(1) Screen collected information;

(2) Resolve all discrepancies and verify that the VEs' certifications are properly completed; and

(3) For qualified examinees, forward electronically all required data to the FCC. All data forwarded must be retained for at least 15 months and must be made available to the FCC upon request.

(c) Each VEC must make any examination records available to the FCC, upon request

(d) The FCC may:

(1) Administer any examination element itself;

(2) Re-administer any examination element previously administered by VEs, either itself or under the supervision of a VEC or VEs designated by the FCC; or

(3) Cancel the operator/primary station license of any licensee who fails to appear for re-administration of an examination when directed by the FCC, or who does not successfully complete any required element that is re-administered. In an instance of such cancellation, the person will be granted an operator/primary station license consistent with completed examination elements that have not been invalidated by not appearing for, or by failing, the examination upon re-administration.

[54 FR 25857, June 20, 1989, as amended at 59 FR 54835, Nov. 2, 1994; 62 FR 17567, Apr. 10, 1997; 63 FR 68981, Dec. 14, 1998; 71 FR 66465, Nov. 15, 2006; 79 FR 35291, June 20, 2014]

§97.521 VEC qualifications.

No organization may serve as a VEC unless it has entered into a written agreement with the FCC. The VEC must abide by the terms of the agreement. In order to be eligible to be a VEC, the entity must:

(a) Be an organization that exists for the purpose of furthering the amateur service;

(b) Be capable of serving as a VEC in at least the VEC region (see appendix 2) proposed;

(c) Agree to coordinate examinations for any class of amateur operator license;

(d) Agree to assure that, for any examination, every examinee qualified under these rules is registered without regard to race, sex, religion, national origin or membership (or lack thereof) in any amateur service organization;

[54 FR 25857, June 20, 1989, as amended at 58 FR 29127, May 19, 1993; 61 FR 9953, Mar. 12, 1996]

§97.523 Question pools.

All VECs must cooperate in maintaining one question pool for each written examination element. Each question pool must contain at least 10 times the number of questions required for a single examination. Each question pool must be published and made available to the public prior to its use for making a question set. Each question on each VEC question pool must be prepared by a VE holding the required FCC-issued operator license. See §97.507(a) of this part.

§97.525 Accrediting VEs.

(a) No VEC may accredit a person as a VE if:

(1) The person does not meet minimum VE statutory qualifications or minimum qualifications as prescribed by this part;

(2) The FCC does not accept the voluntary and uncompensated services of the person;

(3) The VEC determines that the person is not competent to perform the VE functions; or

(4) The VEC determines that questions of the person's integrity or honesty could compromise the examinations.

(b) Each VEC must seek a broad representation of amateur operators to be VEs. No VEC may discriminate in accrediting VEs on the basis of race, sex, religion or national origin; nor on the basis of membership (or lack thereof) in an amateur service organization; nor on the basis of the person accepting or declining to accept reimbursement.

§97.527 Reimbursement for expenses.

VEs and VECs may be reimbursed by examinees for out-of-pocket expenses incurred in preparing, processing, administering, or coordinating an examination for an amateur operator license.

[66 FR 20752, Apr. 25, 2001]

Appendix 1 to Part 97—Places Where the Amateur Service is Regulated by the FCC

In ITU Region 2, the amateur service is regulated by the FCC within the territorial limits of the 50 United States, District of Columbia, Caribbean Insular areas [Commonwealth of Puerto Rico, United States Virgin Islands (50 islets and cays) and Navassa

Island], and Johnston Island (Islets East, Johnston, North and Sand) and Midway Island (Islets Eastern and Sand) in the Pacific Insular areas.

In ITU Region 3, the amateur service is regulated by the FCC within the Pacific Insular territorial limits of American Samoa (seven islands), Baker Island, Commonwealth of Northern Mariana Islands, Guam Island, Howland Island, Jarvis Island, Kingman Reef, Palmyra Island (more than 50 islets) and Wake Island (Islets Peale, Wake and Wilkes).

Appendix 2 to Part 97—VEC Regions

1. Connecticut, Maine, Massachusetts, New Hampshire, Rhode Island and Vermont.

2. New Jersey and New York.

3. Delaware, District of Columbia, Maryland and Pennsylvania.

4. Alabama, Florida, Georgia, Kentucky, North Carolina, South Carolina, Tennessee and Virginia.

5. Arkansas, Louisiana, Mississippi, New Mexico, Oklahoma and Texas.

6. California.

7. Arizona, Idaho, Montana, Nevada, Oregon, Utah, Washington and Wyoming.

8. Michigan, Ohio and West Virginia.

9. Illinois, Indiana and Wisconsin.

10. Colorado, Iowa, Kansas, Minnesota, Missouri, Nebraska, North Dakota and South Dakota.

11. Alaska.

12. Caribbean Insular areas.

13. Hawaii and Pacific Insular areas.

* End of FCC Part 97 *

§2.106, footnote US270

(This footnote is referenced in §97.313)

47 CFR §2.106 Table of Frequency Allocations. United States (US) Footnotes (These footnotes, each consisting of the letters "US" followed by one or more digits, denote stipulations applicable to both Federal and non-Federal operations and thus appear in both the Federal Table and the non-Federal Table.)

US270 In the band 420 - 450 MHz, the following provisions shall apply to the amateur service:

(a) The peak envelope power of an amateur station shall not exceed 50 watts in the following areas, unless expressly authorized by the FCC after mutual agreement, on a case-by-case basis, between the District Director of the applicable field office and the military area frequency coordinator at the applicable military base. For areas (5) through (7), the appropriate military coordinator is located at Peterson AFB, CO.

  (1) Arizona, Florida and New Mexico.

  (2) Within those portions of California and Nevada that are south of latitude 37°0′ N.

  (3) Within that portion of Texas that is west of longitude 104° W.

  (4) Within 322 km of Eglin AFB, FL (30°30′ N, 86°30′ W); Patrick AFB, FL (28°21′ N, 80°43′ W); and the Pacific Missile Test Center, Point Mugu, CA (34°09′ N, 119°11′ W).

  (5) Within 240 km of Beale AFB, CA (39°08′ N, 121°26′ W).

  (6) Within 200 km of Goodfellow AFB, TX (31°25′ N, 100°24′ W) and Warner Robins AFB, GA (32°38′ N, 83°35′ W).

  (7) Within 160 km of Clear AFS, AK (64°17′ N, 149°10′ W); Concrete, ND (48°43′ N, 97°54′ W); and Otis AFB, MA (41°45′ N, 70°32′ W).

(b) In the sub-band 420 - 430 MHz, the amateur service is not allocated north of Line A (def. §2.1).

How to Renew Your License

1. Log in to:

    https://wireless2.fcc.gov/UlsEntry/licManager/login.jsp

    with your FCC Registration Number (FRN) and password.
    Note: If you are taken to the My Applications page, click My Licenses to begin the steps below.

2. On the left-side of the screen, click Renew Licenses.

3. Select the licenses and Click Add.

4. Click Continue to navigate through the application.

5. Sign your application and click Submit Application. *By typing your first and last name in the name fields you have signed your Application.*

If you prefer to renew by mail, you may use FCC Form 605 (edition date of July 2005 or later) to apply for renewal of your amateur operator/primary station license grant, including those station license grants within the grace period. Use purpose RO (Renew Only) for this. If it is necessary to also modify your name or mailing address shown on your amateur operator/primary station license grant, use purpose RM (Renew/Modify). If you are not requesting a call sign change, file the Main form only. (Club or military recreation stations need to file through a Club Station Call Sign Administrator. RACES stations are no longer being renewed.)

Mail the Form 605 to:

FCC
1270 Fairfield Road
Gettysburg, PA 17325 - 7245

You may also deliver the Form 605 in person to the same address.

You may receive a license expiration notification from an entity, but you are not required to use its services. File for renewal 30 - 60 days, but no more than 90 days, before your license expiration date. 47 CFR Section 97.21, provides that when your application for renewal has been received by the FCC on or before the license expiration date, your operating authority is continued until the final disposition of your application. If your license expires, you may apply for renewal of the license for another term during a two-year filing grace period. The application document must be received by the FCC on or before the end of the grace period. Unless and until the license is renewed, no amateur operator or station operating privileges are conferred. Applications received after the grace period has ended cannot be granted.

Section 1.913(f) requires that all applications, other than applications that must be filed through a VEC, may be submitted manually to the address below in number 11.

Until Vanity licenses become renewable, there is no fee for renewing your amateur operator/primary station license grant.

- Licenses cannot be renewed earlier than 90 days before the expiration date.

- Licenses past the expiration date are placed in an "ineligible" list for a period of time. In step 3 above, click the button "Ineligible." If the licenses appear, you can file a renewal application. However, you must also include a waiver request along with a justification for filing the renewal application late.

If you previously started to file a renewal application and did not complete the filing, it could be in a list of Saved Applications.

Note: The address and contact information you have entered in CORES registration will not be automatically associated with your licenses. To change the address or other contact information on your license, you must update your information in ULS or submit Form 605 manually.

How to Apply for a Vanity Call Sign

The FCC offers amateur radio licensees the opportunity to request a specific call sign for a primary station and for a club station. A vanity call sign is selected by the FCC from a list of call signs requested by the station licensee or the club station license trustee. Military recreation stations are not eligible for a vanity call sign. To make the request for a vanity call sign, you can use either the Universal Licensing System or Form 605 Main Form and Schedule D.

Obtaining Vanity Call Signs

You can obtain a vanity call sign using either the Universal Licensing System or FCC Form 605, Main Form and Schedule D.

There are up to three different request types for requesting vanity call signs, depending on whether the call sign is to apply to a primary or a club station. The requestor may list up to twenty-five call signs in order of preference. The exact prefix, numeral, and suffix must be given for each call sign. Requests stated in general terms such as, "Any call sign with my initials" or "Any call sign having the prefix (or suffix) _____" will be dismissed. The first assignable call sign on the list for which the requestor is eligible will be shown on the license grant for the requestor's station and the original call sign will be vacated.

For most amateurs, the process is;

a) Create a list of up to twenty-five of your favorite call signs, with your number one pick at the top of the list.

b) Submit the list to the FCC. This process can be accomplished on the ULS web site, by logging in using your FRN and password.

c) The FCC will go down the list from top to bottom. The first call sign on your list that is available is the one you'll get.

Objections to the assignment of call signs requested by another licensee or a club station trustee will not be entertained at the FCC. However, this does not hamper any party from asserting such rights as it may have under private law in some other forum. The FCC does not consider an individual to be a former holder where the call sign was originally obtained through bribery, fraud, favoritism or other improper means. A club station trustee of a club that has been assigned a vanity call sign is not eligible to apply for any additional vanity call signs.

Note: You cannot receive a vanity call sign from a call sign group for which your operator class is not eligible. For example, if you are operator class T (technician), you can only receive call signs from groups C & D. If you request a call sign from groups A or B, your application will be dismissed by the Commission.

Call Sign Availability

A call sign is normally assignable two years following license expiration, surrender, revocation, set aside, cancellation, void *ab initio*, or death of the grantee. Where a vanity call sign for which the most recent recipient was ineligible is surrendered, cancelled, revoked or voided, the call sign is not assignable for 30 days following the date such action is taken, or for the period for which the call sign would not have been available to the vanity call sign system pursuant to Section 97.19(c)(2) or (3) but for the intervening grant to the ineligible applicant, whichever is later. Refer to the Sequential Call Sign System for how call signs are sequentially assigned and the grouping and geographic region attributes of each call sign.

The following call signs are not available for assignment:

1) KA2AA-KA9ZZ, KC4AAA-KC4AAF, KC4USA-KC4USZ, KG4AA-KG4ZZ, KC6AA-KC6ZZ, KL9KAA- KL9KHZ, KX6AA-KX6ZZ;

2) Any call sign having the letters SOS or QRA-QUZ as the suffix;

3) Any call sign having the letters AM-AZ as the prefix (these prefixes are assigned to other countries by the ITU);

4) Any 2-by-3 format call sign having the letter X as the first letter of the suffix;

5) Any 2-by-3 format call sign having the letters AF, KF, NF, or WF as the prefix and the letters EMA as the suffix (U.S Government FEMA stations);

6) Any 2-by-3 format call sign having the letters AA-AL as the prefix;

7) Any 2-by-3 format call sign having the letters NA-NZ as the prefix;

8) Any 2-by-3 format call sign having the letters WC, WK, WM, WR, or WT as the prefix (Group X call signs);

9) Any 2-by-3 format call sign having the letters KP, NP or WP as the prefix and the numeral 0, 6, 7, 8 or 9;

10) Any 2-by-2 format call sign having the letters KP, NP or WP as the prefix and the numeral 0, 6, 7, 8 or 9;

11) Any 2-by-1 format call sign having the letters KP, NP or WP as the prefix and the numeral 0, 6, 7, 8 or 9;

12) Call signs having the single letter prefix (K, N or W), a single digit numeral 0, 1, 2, 3, 4, 5, 6, 7, 8, 9 and a single letter suffix are reserved for the special event call sign system.

General Rules

You must hold an unexpired amateur operator/primary station license grant of the proper operator class, as described below, to request a vanity call sign for your primary station.

To request a vanity call sign for a club station, you must also hold an unexpired club station license grant listing you as the license trustee.

Your name and mailing address as shown on your current license grant must be correct.

If your license grant has expired, you must first renew the license. After the renewal of your license is granted, you may file a vanity call sign request. You can change your name or address at the time of your vanity call sign request, by filing FCC Form 605.

The call sign you are requesting may already be assigned. Refer to the Universal Licensing System License Search for verification.

The license of the former holder now deceased must show a status of expired or cancelled in the licensee database. See Section 97.31(a). This is accomplished by submitting a request that includes a death certificate or obituary that shows the person named in the operator/primary station license grant has died. Such a request may be submitted as a pleading associated with the deceased licensee's license and sent to:

FCC
1270 Fairfield Road
Gettysburg, PA 17325 - 7245

In addition, the Commission may cancel a call sign if it becomes aware of the grantee's death through other means. No action will be taken during the last thirty days of the post-expiration grace period on a request to cancel a call sign due to the grantee's death.

The information for cancellation of a call sign must be submitted prior to filing the vanity application.

As of 2015, there is no longer any fee for a vanity call sign.

Amateur Radio Web Links

Note: All links were tested and live shortly before publication of this year's edition of the Book of Facts. No endorsement or criticism is implied by the inclusion or exclusion of any link listed here.

Category	Name	Link
Amateur Radio News	Amateur Radio Newsline	http://www.arnewsline.org/
	ARRL Amateur Radio News	http://www.arrl.org/
	CQ Website News page	http://newsvc.cq-amateur-radio.com/
	Ham Nation	https://www.youtube.com/user/TWiTHamNation
	KB6NU's Ham Radio Blog	https://www.kb6nu.com/
	QRZ.com	http://www.qrz.com/
Amateur Radio Organizations	ARRL	http://www.arrl.org/
	ARRL - Find a club	http://www.arrl.org/find-a-club
	ARRL - Find a hamfest	http://www.arrl.org/hamfests/search
	ARRL - Find an exam	http://www.arrl.org/find-an-amateur-radio-license-exam-session
	Deutscher Amateur-Radio-Club - DARC (English language)	http://www.darc.de/ausland/new/englisch/index.html
	FISTS Int'l Morse Preservation Society	http://www.fists.org/
	IARU Region 1 (Europe/Africa)	http://www.iaru-r1.org/
	IARU Region 2 (N./S. America)	http://www.iaru-r2.org/

Category	Name	Link
	IARU Region 3 (Asia/Oceania)	http://www.jarl.or.jp/iaru-r3
	International Amateur Radio Union - IARU	http://www.iaru.org/
	Irish Radio Transmitters Society - IRTS	http://www.irts.ie/
	New Zealand Association of Radio Transmitters - NZART	http://www.nzart.org.nz/nzart/
	Quarter Century Wireless Assn. - QCWA	http://www.qcwa.org/
	Radio Amateur Satellite Corp. - AMSAT	http://www.amsat.org/
	Radio Amateurs of Canada - RAC	http://www.rac.ca/
	Radio Society of Great Britain - RSGB	http://www.rsgb.org.uk/
	Ten-X International Net - 10 - 10	http://www.ten-ten.org/
	Tucson Amateur Packet Radio - TAPR	http://www.tapr.org/
	YLRL - YL Radio League	https://www.ylrl.org/
Amateur Radio Software	Log4OM	http://www.log4om.com/
	Aatest	http://www.dxsoft.com/en/products/aalog_contest/
	AC Log by N3FJP	http://www.n3fjp.com/
	Airlink Express	http://www.airlinkexpress.org/
	Black Cat Systems (multiple software products)	http://www.blackcatsystems.com/
	CallSign Software - radio control software	http://www.callsignsoftware.com/
	CHIRP	https://chirp.danplanet.com/projects/chirp/wiki/Home
	CommCat - logging and radio control for iPhone, iPad	http://www.commcat.com/
	DigiPan	http://www.digipan.net/
	DX Atlas	http://www.dxatlas.com/
	DX Toolbox	https://www.blackcatsystems.com/software/ham-shortwave-radio-propagation-software.html
	DXLab Suite	https://www.dxlabsuite.com/
	Echolink	http://www.echolink.org/
	e-Ham reviews of logging software	https://www.eham.net/reviews/products/27
	EZNEC Antenna Modeling	https://www.eznec.com/
	fldigi (Multiple digital mode software)	https://sourceforge.net/projects/fldigi/
	Ham Radio Deluxe	https://www.hamradiodeluxe.com/
	HamSphere	https://www.hamsphere.com/
	iCluster - DX cluster software for iPhone	http://ham-radio-apps.com/icluster/

Category	Name	Link
	Logbook of The World (LoTW)	http://www.arrl.org/lotw/
	Logger 32	http://www.logger32.net/
	MacLoggerDX	http://www.dogparksoftware.com/MacLoggerDX.html
	MMSSTV	https://hamsoft.ca/pages/mmsstv.php
	MMTTY	https://hamsoft.ca/pages/mmtty.php
	MultiPSK	http://f6cte.free.fr/index_anglais.htm
	N1MMlogger	http://www.n1mm.com/
	N3JFP Contesting Software	http://www.n3fjp.com/
	Open Repeater Project - Linux based open source repeater controller	https://openrepeater.com/
	Overview of HF digital mode software	http://www.hamuniverse.com/hfdigitalmodessoftware.html
	POLAR Electric - Morse code decoding software	http://www.polar-electric.com/
	Prolog	http://www.prologsystem.com/
	Proplab-Pro - propagation software	http://www.spacew.com/www/proplab.html
	RF Toolbox - app for iPhone	http://electronic-toolbox.com/rf-toolbox/
	RT Systems	https://www.rtsystemsinc.com/default.asp
	TrueTTY	http://www.dxsoft.com/en/products/truetty/
	WSJT Suite	https://physics.princeton.edu/pulsar/k1jt/
	WSPR (Weak Signal Propagation Reporter)	http://physics.princeton.edu/pulsar/K1JT/wspr.html
Call Sign Servers	ARRL	http://www.arrl.org
	Ham Call/Buckmaster	http://hamcall.net/
	QRZ.com	http://www.qrz.com/
	Univ. of Arkansas / Little Rock	http://callsign.ualr.edu/callsign.shtml
	Vanity call sign info	http://www.vanityhq.com/
Dealers	Associated Radio	http://www.associatedradio.com/home.php
	B&H Sales, Inc.	http://www.hamradiocenter.biz/
	Cheapham.com	https://www.cheapham.com/
	DX Engineering	https://www.dxengineering.com/
	DX Store (Misc.)	http://www.dxstore.com/
	Gigaparts	https://www.gigaparts.com/
	Ham Radio Outlet	http://hamradio.com

Category	Name	Link
	HamPROs!	http://www.hampros.com/
	MFJ Enterprises	http://www.mfjenterprises.com/
	MTC	https://www.mtcradio.com/
	Universal Radio, Inc.	http://www.universal-radio.com/
Education	Code Quick	http://www.cq2k.com/index.html
	Command Production - FCC General Radiotelephone License Training	http://www.licensetraining.com/
	Fast Track Ham License Programs, ham radio books, videos	http://fasttrackham.com
	Radio-Electronics.com	https://www.radio-electronics.com/
	W5YI - VEC, VE, FCC Commercial license materials, amateur license study	http://w5yi.com/
Equipment Reviews	e-Ham.net	http://www.eham.net
Equipment, parts & supplies	A Ham Radio Auction	http://www.ahamradioauction.com/
	ABR Industries	http://www.abrind.com/
	ACOM International	http://www.hfpower.com/
	Alltronics	http://www.alltronics.com/
	Alpha Antenna	http://www.alphaantenna.com/
	Alpha Delta Communications	https://www.alphadeltaradio.com/
	Aluma Towers	http://www.alumatower.com/
	American Radio Supply	https://www.americanradiosupply.com/
	Ameritron	http://www.ameritron.com/
	Amidon Corp.	http://www.amidoncorp.com/
	Amidon Inductor Cores	http://www.amidoncorp.com/
	Antennas US	https://www.antennas.us/
	Antique Electronic Supply	http://www.tubesandmore.com/
	AOR, Inc.	http://www.aorusa.com/
	Arcom Controllers - repeater controllers	http://www.arcomcontrollers.com/
	Arlan Communications	http://www.arlancommunications.com/
	Array Solutions, Inc.	https://www.arraysolutions.com/
	ASA, Inc. (Waterproof log books)	http://www.waterprooflogbooks.com/
	AVVid (Amateur Radio Service Center)	http://www.avvid.com/
	Balun Designs	https://www.balundesigns.com/

Category	Name	Link
	Barker & Williamson Antennas	http://www.bwantennas.com/
	Batteries America	http://www.batteriesamerica.com/
	Begali Keys	http://www.i2rtf.com/
	Bencher Keys	http://www.vibroplex.com/
	bhi Ltd.	https://www.bhi-ltd.com/
	Bilal Co./Isotron Antennas	http://www.isotronantennas.com/
	Bozak Antennas	http://bozak.tripod.com/HOMEPAGE.HTM
	BridgeCom Systems	https://www.bridgecomsystems.com/
	Buddipole Antennas	http://www.buddipole.com/
	Butternut Antennas	http://www.dxengineering.com
	C.A.T.S. (Rotor parts)	http://www.rotor-parts.com/
	Cable X-Perts, Inc.	http://www.cablexperts.com/
	CAL-AV	http://www.cal-av.com/
	Carl's Electronics	https://www.electronickits.com/
	Chameleon Antennas	https://chameleonantenna.com/
	Champion Radio Products (Towers)	https://www.championradio.com/
	Charlie's Electronics - Amateur Radio Repairs	http://www.charlieselectronics.com/
	Circuit Specialists	https://www.circuitspecialists.com/
	Clear Signal Products, Inc.	http://www.coaxman.com/
	Comet Antennas/NCG meters, etc.	http://www.natcommgroup.com/
	Communication Concepts, Inc.	http://www.communication-concepts.com/
	Communications Service Company - antennas & towers	https://www.kj6y.com/
	Compudigital Industries - vintage Kenwood renovation	https://k6iok.com/
	Computer 2100, LLC - Monitoring and analysis tools and hardware accessories	https://www.computer2100.com/
	ComTek Systems	https://www.dxengineering.com/search/brand/comtek
	Critical Towers	http://www.criticaltowers.com/
	Cutting Edge Enterprises - Portable power	http://www.powerportstore.com/
	Daiwa	http://www.natcommgroup.com/
	Davis Instruments	https://www.davisinstruments.com/
	Davis RF	http://www.davisrf.com/
	Diamond Antennas	https://www.rfparts.com/antenna/antenna-

Category	Name	Link
		diamond.html
	Dishtronix, Inc. - Power amps	http://www.dishtronix.com/
	DZ Company (Kits)	http://www.dzkit.com/
	Electronic Products Design, Inc. - One-off, small run, large run electronics mfgr.	http://www.epd-inc.com/
	Electronics USA - LED digital timers, clocks, etc., code keys	http://electronicsusa.com/
	Elk Antennas - Log periodic antennas	https://elkantennas.com/
	Expert Linears - power amps	http://expertlinears.com/
	EZ Hang, Inc. - Antenna launcher	http://www.ezhang.com/
	Fractal Antenna Systems	http://www.fractenna.com/
	Front Panel Express, LLC - Custom front panels	http://www.frontpanelexpress.com/
	GAP Antenna Products	http://www.gapantenna.com/
	Geochron, Inc. - "Moving map" clocks	http://www.geochronusa.com/
	Gifts4Hams	http://www.gifts4hams.com/
	Green Heron Engineering LLC - Antenna rotators	http://www.greenheronengineering.com/
	Hagerty Radio Company	http://www.wa1ffl.com/
	Ham Services - tower & antenna service	http://www.hamservices.com/
	Heil Sound	https://heilsound.com/
	Hy Power Antenna Company	http://www.angelfire.com/electronic/hypower/
	Hy-Gain Antennas	http://www.hygain.com/
	ICS Controllers	https://ics-ctrl.com/
	InnoVAntennas	https://www.innovantennas.com/index.php?lang=en
	International Radio Products - INRAD - Crystal filters	http://www.inrad.net/
	Isotron Antennas	http://www.isotronantennas.com/
	Jetstream	http://www.jetstream-usa.com/
	JK Antennas	https://jkantennas.com/
	K4AVU Amateur Radio Products	http://www.k4avu.webs.com/
	KB9VBR Antennas	https://www.jpole-antenna.com/
	Larry's Antennas	http://www.kj7u.com/
	LDG Electronics - Antenna tuners	https://ldgelectronics.com/
	Lightning Prevention Systems, Inc	https://www.lpsnet.com/
	M2 Antenna Systems, Inc.	https://www.m2inc.com/

Category	Name	Link
	Mastrant - Antenna guying	http://www.mastrant.com/
	Mitchell Electronics - Transformers	https://www.yesmec.com/
	Morse Express	http://www.morsex.com/
	N3ZN Keys, LLC	http://www.n3znkeys.com/
	National RF, Inc.	http://www.nationalrf.com/
	Nemal Electronics - cables & connectors, misc.	http://www.nemal.com/
	Norms Fabrication - Tower parts	http://www.normsfab.com/
	Old Heathkit Parts	http://www.oldheathkitparts.com/
	OLYMPIX - RFI Filters	https://olympixcorp.com/rfchoke/index.htm
	Online Cables (Applied Interconnect)	https://www.onlinecables.com/legacy/
	Palomar Engineers - RFI solutions	https://palomar-engineers.com/
	Palstar - amps and antenna tuners	http://www.palstar.com/en/la1k/
	Peter W. Dahl - custom transformers & chokes	http://www.pwdahl.com/
	Pixel Satellite Radio - GHz range antennas & accessories	http://www.pixelsatradio.com/
	Polyphaser Corporation - Surge & lightning protection	https://www.polyphaser.com/
	PowerWerx - Powerpoles®	http://www.powerwerx.com/
	preciseRF - Precision measurement instruments & mag loop antennas	http://preciserf.com/
	Precision Resistors	http://www.riedon.com/
	QRO Technologies, Inc.	http://www.qrotec.com/
	QTH.com Classified Ads	https://swap.qth.com/
	R & L Electronics, Inc	http://www.randl.com/
	Rescue Tape	https://www.rescuetape.com/
	RF Connection	https://rfconnection.com/
	RF Gear 2 Go	https://www.rfgear2go.com/
	RF Parts Company	http://www.rfparts.com/
	RigExpert - Antenna analyzers	https://rigexpert.com/
	Rotating Tower Systems	http://www.custommetalworks.com/rts/rts.htm
	Solder-it - Butane powered soldering irons	http://www.solder-it.com/
	SteppIR Antennas Inc.	http://www.steppir.com/
	Super Antenna	http://newsuperantenna.com/
	Tarheel Antennas, Inc.	http://www.tarheelantennas.com/

Category	Name	Link
	Tennadyne, L.L.C. - Log periodic antennas	http://www.tennadyne.com/
	Ten-Tec, Inc.	http://www.tentec.com/
	Timewave Technology - voice and digital communications products	http://www.timewave.com/
	Tower building book	https://www.championradio.com/UP-THE-TOWER-The-Complete-Guide-To-Tower-Construction.1
	Uniden America Corporation	http://www.uniden.com/
	US Tower Corporation	http://www.ustower.com/
	Ventenna - portable and stealth antennas	https://www.ventenna.com/
	Vibroplex	http://www.vibroplex.com/
	W&W Manufacturing Co. - Radio batteries and accessories	http://www.ww-manufacturing.com/
	W2IHY Technologies - Audio and RF switching and processing	http://www.w2ihy.com/
	WaveNode SWR/RF/DC Metering	https://wavenodedevelop.com/
	WD7S Productions - HF Amp Controls	http://home.earthlink.net/~wd7s/
	Weather Defender - Weather tracking software	http://www.weatherdefender.com/
	West Mountain Radio	http://www.westmountainradio.com/
	Western Case - fitted carrying cases for small electronics	http://www.westerncasecompany.com/
	Wired Communications - connectors, adapters, coax	https://www.wiredco.com/
	Wireman	http://www.thewireman.com/
	WK5R Circuit Boards	http://www.wk5r.org/
	Xmatch Antenna Tuner	http://n4xm.myiglou.com/
Government Agencies & Resources	Amateur Radio on the International Space Station - ARISS	http://www.rac.ca/ariss/
	Electromagnetic spectrum chart	http://www.its.bldrdoc.gov/fs-1037/images/frqcharc.gif
	FCC	http://www.fcc.gov/
	FCC Amateur Radio Information Page	http://www.cq-amateur-radio.com/cq_hobby_radio_links/wireless.fcc.gov/services/amateur/
	FCC Universal Licensing System - ULS	http://wireless.fcc.gov/uls
	FCC Wireless Telecommunications Bureau	http://wireless.fcc.gov/
	Industry Canada	http://www.ic.gc.ca/
	Industry Canada Amateur Radio page	http://www.ic.gc.ca/epic/site/smt-gst.nsf/en/h_sf01709e.html

Category	Name	Link
	Institute for Telecommunications Services- ITS	http://www.its.bldrdoc.gov/
	International Telecommunications Union - ITU	http://www.itu.int/
	National Aeronautics and Space Admininstration - NASA	http://www.nasa.gov/
	National Hurricane Center	http://www.nhc.noaa.gov/
	National Institute of Standards & Technology - NIST	http://www.nist.gov/
	National Oceanic & Atmospheric Administration - NOAA	http://www.noaa.gov/
	National Telecommunications and Information Administration - NTIA	http://www.ntia.doc.gov/
	National Weather Service	http://www.nws.noaa.gov/
	NIST Official Time	http://nist.time.gov/
	NIST Time & Frequency Division - WWV	http://www.boulder.nist.gov/timefreq/index.html
	NOAA Storm Prediction Center	http://www.spc.noaa.gov/
	RF Spectrum Allocation Chart PDF	http://www.ntia.doc.gov/osmhome/allochrt.pdf
	US Census Bureau - printable county maps of each state	https://www.census.gov/geo/maps-data/maps/stcou_outline.html
Miscellaneous	Capital Engraving	http://www.scattercreek.com/~capengrave/
	Cheap QSL's	https://www.cheapqsls.com/
	CircuitsArchive - archive of schematics	https://circuitsarchive.org/
	Daily DX - daily e-mail DX newsletter	http://www.dailydx.com/
	eQSLCard - Free web-based QSL card distribution	https://www.eqsl.cc/QSLCard/Index.cfm
	Ham Gear (Name tags & embroidery)	https://squareup.com/store/e-and-m-embroidery
	Ham Radio Manuals	https://www.hamradiomanuals.com/
	Hip-Ham T-Shirts	https://www.hiphamshirts.com/
	ManualMan - vintage radio-related operating manuals	http://www.manualman.com/
	NYQSL	http://www.nyqsl.com/
	Personalized Map Company - online maps	http://www.mymaps.com/hamorder.htm
	Photo QSL's	http://www.photoqsls.com/
	ProPrints - printing for Hams	http://www.proprints.com/hamgear.html
	Vintage Manuals, Inc.	http://www.vintagemanuals.com/
Propagation Information	DX Maps	https://www.dxmaps.com/spots/mapg.php
	hfradio.com	http://prop.hfradio.org/
	N0BNH - hamqsl.com	http://www.hamqsl.com

Category	Name	Link
	NOAA Space Environment Center	http://www.sec.noaa.gov/
	SpaceWeather	http://www.spaceweather.com/
	WSPRNET	http://wsprnet.org/drupal/wsprnet/map
Radio Manufacturers	Alinco Electronics	http://www.alinco.com
	ELAD SDR	http://shop.elad-usa.com/
	Elecraft - Transceivers & accessories	http://www.elecraft.com/
	FlexRadio Systems	http://www.flex-radio.com/
	ICOM America	http://www.icomamerica.com/
	Kenwood	http://www.kenwoodusa.com/
	Yaesu USA	http://yaesu.com/
Repeater Listings	Repeaterbook	http://repeaterbook.com

Amateur Radio Glossary

Term	Definition
0 – 9	
10 codes	Prosigns -- over-the-air shorthand -- used by emergency responders and Citizens Band hobbyists. Not generally well received nor broadly comprehended in the amateur radio community.
72	73 in QRP operation.
73	Prosign for "best regards" or "best wishes."
88	Prosign for "love and kisses."
A	
A.R.C.	Amateur Radio Club. For instance, the Mount Diablo Amateur Radio Club is known to its members and friends as MDARC -- EMM-dark.
AC (alternating current)	An electrical current which flows first one direction then the opposite and, typically, continues to do so.
Adaptive filter	A digital filter used in digital signal processing.
ADC	Analog to Digital Converter. In amateur radio, most commonly used to turn audio or RF into digital data.

Term	Definition
Adjacent-channel interference	Interference received from a nearby frequency.
Admittance	The reciprocal of impedance. Measured in siemens.
AF	Audio Frequency.
AFC	Automatic Frequency Control. A circuit that automatically compensates for frequency drift.
AFSK	Audio Frequency Shift Keying. An analog modulation method for transferring digital data such as RTTY text.
AGC	Automatic Gain Control. A circuit that automatically optimizes receiver gain.
AGL	Above Ground Level, for antenna installations.
Airlink Express	Multiple digital HF mode software.
ALC	Automatic Level Control. A circuit that automatically limits RF drive to the final amplifier to prevent distortion.
AM	Amplitude modulation. While SSB is a form of amplitude modulation, in amateur radio use, AM is normally double sideband modulation with a full carrier.
Amateur operator	A person holding a written authorization to be the control operator of an amateur station.
Amateur Service	"A radiocommunication service for the purpose of self-training, intercommunication and technical investigations carried out by amateurs, that is, duly authorized persons interested in radio technique solely with a personal aim and without pecuniary intere
Amateur station	A station licensed in the amateur service, including necessary equipment, used for amateur communication.
Ammeter	A test instrument that measures the amperage passing through a circuit.
Amperage	The amount of current flowing through a conductor or circuit.
Ampere (A)	The unit of current flow. 1 ampere = 1 coulomb of charge passing a point in 1 second.
Amplitude Modulation	A system of transmitting information by modulating the amplitude of an RF carrier wave.
AMSAT	Amateur radio SATellite. The name for amateur radio satellite organizations world-wide, but in particular the Radio Amateur Satellite Corporation.
Term	Definition

Term	Definition
AMTOR	Amateur Teleprinting Over Radio. A form of RTTY.
Analog (communication)	The transmission of signals or information represented via a continuously variable dimension, such as amplitude, frequency, or phase.
ANF	Automatic Notch Filter.
Antenna	A wire, rod, pole, or other device for radiating or receiving electromagnetic waves.
Antenna analyzer	An instrument used to measure and display various characteristics of antennas at various frequencies.
Antenna gain	See "gain"
Antenna matching network	See "Antenna tuner"
Antenna party	A long-standing tradition among hams where several gather to assist a fellow ham in mounting antennas and/or towers.
Antenna switch	A specialized switch used to connect one transmitter, receiver, or transceiver to more than one antenna while connecting the other antenna(s) to ground for safety.
Antenna tuner	A circuit or device for matching the impedance of an antenna system to the impedance of the transmitter and/or receiver.
AOS	Acquisition of Signal from a satellite.
APC	Automatic Power Control. Limits current to final amplifier when SWR is high.
Apogee	The point in an orbiting object's orbit where it is farthest from the object around which is it orbiting.
APRS	Automatic Packet Reporting System. APRS combined with a GPS can provide real-time updates of a mobile or handheld transmitter's location.
ARC	Amateur Radio Club. For instance, the Mount Diablo Amateur Radio Club is known to its members and friends as MDARC -- "EMM-dark."
ARES	The Amateur Radio Emergency Service.
ARISS	Amateur Radio on the International Space Station.
ARRL	The American Radio Relay League. The national amateur radio club and lobbying organization - our voice at the FCC and Congress.
Term	Definition

Term	Definition
ASCII	American national Standard Code for Information Interchange. A 7 bit digital code for the transmission of text.
ATV	Amateur Television.
Audio frequency (signal)	An AC electrical signal with a frequency in the range of 20 Hz to 20 kilohertz.
Auroral propagation	Propagation from auroras.
Autopatch	A system for connecting a repeater station to the telephone system for the purpose of allowing amateur operators to place phone calls via their radios. (Seldom heard or even existent these days.)
AWG	American Wire Gauge. A system of specifying the diameter of wire.
Azimuth	In the context of directional antennas, a direction in the horizontal plane. Most often shown on a graph of polar coordinates, with 0 degrees corresponding to the far right side of the circular polar graph and the degree scale running counter-clockwise a
B	
Backscatter	Radio signals bounced back toward the transmitter (and elsewhere) from, among other possibilities, ionized patches in the ionosphere, from a rain front, or from aurora.
Balanced line	A feed line with neither conductor at ground potential, such as ladder line.
Balun	A device for connecting an unbalanced source (such as coaxial cable transmission line) to a balanced load (such as a dipole antenna.) A balun can also be a choke balun, used to prevent unwanted RF from traveling back down a transmission line into the shac
Band	A range of frequencies.
Band spread	A receiver specification denoting how far apart on the dial stations on nearby frequencies will seem to be. Somewhat like an RF analog of "zoom." Usually expressed as "kHz per turn" of the tuning knob.
Bandpass filter	A circuit or device that allows a desired band of frequencies to pass while blocking all others.
bands are dead, The	"Communicating on HF using the ionosphere is more difficult today than I had anticipated."
Bandstop filter	A circuit or device that stops an undesired band of frequencies from passing while passing all other frequencies.
Bandwidth	The difference, in Hz, between the lowest and highest frequencies occupied by a given signal, or generated, passed, or successfully processed by a particular device. Technically, the width of a frequency band outside of which the mean power is attenuated

Term	Definition
Baofeng	Amateur radio equipment manufacturer, specializing in inexpensive handhelds and mobile radios. Chinese for "thunder." Pronounced BAO (like "take a bow") fung.
Battery	A device that supplies power by converting chemical energy into electrical energy.
Baud rate	A measure of the rate at which digital data is transferred. Equal to the maximum number of "symbols" that can be sent per second. Not equivalent to bit rate, nor to the number of text characters that can be sent per second. (See "symbol" for more detail.)
Beacon station	An amateur station transmitting for the sole purpose of observation of propagation and reception or other related activities. See NCDXF.
Beam antenna	Any directional antenna.
Beverage antenna	A very long wire antenna, normally for receiving only, mounted near the ground and terminated in a resistor to ground.
BFO	Beat Frequency Oscillator. An oscillator which supplies a signal to a receiver's detector circuit. Necessary to produce an audio tone for CW signals and for SSB.
Bipolar transistor	A type of transistor consisting of a layer of P or N type material sandwiched between two layers of the opposite type of material.
Bird	Nickname for an amateur radio satellite. Also the brand name of a high-end professional grade directional wattmeter.
BNC	Bayonet Neill-Concelman. A type of antenna connector. Also used for some test instruments such as oscilloscopes.
Boat anchor	A large, weighty piece of vintage amateur radio equipment.
BPS	Bits Per Second. A measure of the rate at which digital data is transferred.
BPSK	Binary Phase Shift Keying. A digital data modulation method.
Brass pounder	A CW operator who prefers a straight, mechanical key vs. a "bug" or electronic keyer.
Broadcasting	Transmissions intended to be received by the general public, either direct or relayed.
Bug	A mechanical device for generating the dots and dashes of Morse code semi-automatically.
Bunny hunt	Radiosport of finding hidden transmitters. See also "foxhunting."
Busy lockout	A feature on some transceivers that prevents transmit on a frequency already in use.

Term	Definition
C	
C4FM	The digital data modulation system used by Yaesu's Fusion digital voice and data system.
Call sign	A unique identifier assigned to someone who has attained an amateur radio license.
Capacitance	The amount of electrostatic storage available in a capacitor. Measured in Farads.
Capacitance hat	A system of wires or other structure at the top of a vertical antenna that reduces its inductance and increases its bandwidth.
Capacitor	A component that stores energy in an electrostatic field.
Capture effect	An effect exhibited by FM receivers when a stronger signal "captures" the receiver, blanking another signal.
Card Checker	A volunteer who validates QSL's for those applying for various QSL-based awards.
Carrier	The radio frequency sine wave that is modulated with the information to be transmitted.
Carrier frequency	The center frequency of the band of frequencies in an information carrying signal being transmitted or received.
Cartesian coordinates	See "rectangular coordinates."
Cavity filter	A type of very narrow-band RF filter. Typically found in repeater installations as part of the "duplexer."
Center-fed	Applied to antennas with a feed point at the center of two halves of the antenna. The "+" output of the transmitter connects to one half, the "-" to the other half, as in a typical wire dipole.
Chassis ground	A common connection for all components in a device that connect to the negative side of the power supply.
CHIRP	Free software for programming a wide array of transceivers with a computer. From http://danplanet.com
Chirp	In CW, a slight shift in the transmitter frequency each time the key is closed, which creates a characteristic "chirping" sound at the receiver.
Choke	A inductor used to block alternating or pulsed direct current, sometimes while passing steady direct current. One might use a "choke" on the power supply to a mobile radio to allow 12 V DC to pass while blocking intermittent noise pulses from the ignition

Term	Definition
Circulator	A device in which an RF signal entering any port is transmitted to the next port only. Common in repeater installations.
Clarifier	Alternate name for Receiver Incremental Tuning control.
Clipping	Waveform created when an amplifier is overdriven.
Closed repeater	A repeater that can only be used by those in possession of a special code.
CLOVER	A full-duplex Phase Shift Keying data mode.
CNDX	Conditions.
Coax	Coaxial cable. Transmission line with one conductor inside another, the two being separated by an insulating layer known as the dielectric.
Code key	A device for sending Morse code.
Color code	In the context of transmission and reception, the form of modulation used to transmit information. For instance, CW, SSB, Phase Shift Keying, etc.
Common mode current	In a two (or more) conductor cable; conductor currents not matched by exactly opposite and equal magnitude currents. Common mode currents can lead to excessive noise, RF interference, and even RF burns from equipment in the shack. (See also differential c
Communications emergency	A situation in which communication is required in connection with the immediate safety of human life and/or immediate protection of property.
Complete circuit	An electrical circuit that presents a continuous path for the flow of current.
Condenser	Vintage term for a capacitor.
Conductivity	The reciprocal of resistance. Measured in siemens.
Conductor	A material that will readily pass an electric current.
Contesting	Creating contacts with as many stations as possible over a specific amount of time, and often on specific bands.
Control operator	The party responsible for the transmissions of a station.
Control point	The point at which the control operator function is performed.

Term	Definition
Controlled environment (RF exposure)	An area where an RF signal may cause radiation exposure to people who are aware of the radiation and can exercise some control over their exposure. There are different RF exposure standards for controlled environments vs. uncontrolled environments.
Conventional current	The flow of electrical charge from a positively charged location to a negatively charged location. Until the discovery of the electron, it was thought electricity flowed from positive to negative, so all schematic diagrams use this convention. It is perfe
Coronal hole	Sunspot related solar activity that may enhance 10 meter and VHF propagation while degrading HF propagation.
Coulomb	The basic unit of electrical charge. 1 Coulomb = 6.25 × 1018 electrons.
Counterpoise	Precisely, a ground plane radial or radials mounted above the surface, but used more generally for any ground plane.
Courtesy tone	A repeater feature that causes a tone or beep to be transmitted at the end of each transmission.
Coverage area	The area in which a station can be received.
CQ	Prosign for "Calling anyone." Can be used in CW or phone transmissions. Primarily used on HF and VHF/UHF simplex -- not commonly used on repeaters.
CQ Magazine	One of the two major general interest amateur radio magazines, the other being QST Magazine.
Cross modulation	Interference caused by the interaction of two (or more) carriers.
Crossband repeat	A mode available on some dual-band VHF/UHF radios allowing the radio to receive on one band and simultaneously transmit on the other.
Crystal	A crystal of silicon machined to resonate at a specific frequency when an electric current is applied to it.
CTCSS	Continuous Tone Coded Squelch System. A system of sub-audible tones used to control most repeaters. When the right tone is added to your transmission on the proper frequency, the tone "opens up" the repeater.
Current	The flow of electricity through a conductor.
CW	Continuous Wave. Transmission of information, almost universally by Morse code, by switching the carrier off an on. Colloquially, in amateur radio use, any Morse code transmission, or the code itself.
CW filter	A circuit in a receiver that limits the width of the IF passband to improve selectivity in crowded band conditions.

Term	Definition
D	
D region	The lowest layer of the ionosphere, present only during daylight hours.
DAC	Digital to Analog Converter. In amateur radio, most commonly used to turn data into audio.
dB	Decibels. Not a unit of measure, but rather a logarithmic expression of the ratio between two values. Can be suffixed with various letters to indicate the basis of comparison. For instance, dBm is "decibels of difference compared to a 1 mW signal."0
dBd	The gain of an antenna relative to that of a 1/2 wave dipole antenna.
dBi	The gain of an antenna relative to an isotropic antenna.
DC (direct current)	An electrical current which flows only one direction. It may vary in intensity or even be intermittent, but it still flows in only one direction.
DCS	Digital Coded Squelch. An alternative to CTCSS for repeater access control.
Detector	A circuit in a receiver that retrieves the information from the received signal.
Deviation	A measure of the maximum frequency changes on either side of the carrier center frequency for an FM signal.
DF	Direction finding. Locating a source of a radio transmission through the use of directional antennas and associated equipment.
Differential current	In a two (or more) conductor cable; equal currents flowing in opposite directions in the two conductors. (See also, common mode current.)
DigiPan	HF digital mode software.
Digipeater	A packet radio station used to relay digital packet messages.
Digital (communications)	Computer-based amateur radio modes. Can be data modes, such as packet or APRS, or text based modes such as RTTY or the WSJT modes.
Digital Radio Mondiale. A digital mode designed to compress signals into reduced bandwidth. Not to be confused with DMR, a digital voice mode.	Digital Radio Mondiale. A digital mode designed to compress signals into reduced bandwidth. Not to be confused with DMR, a digital voice mode.

Term	Definition
Diode	A solid-state device that acts as a one-way valve for electricity, among other applications.
Diplexer	A device for frequency-domain multiplexing. A diplexer allows two transmitters to use the same antenna on different frequencies simultaneously.
Directional wattmeter	See "wattmeter"
Director element	A passive antenna element, typically located in front of the driven element.
Dish	A highly direction parabolic dish antenna.
DMR	Digital Mobile Radio. A digital voice mode.
Domino EX	A digital mode that uses a modified form of MFSK known as IFK, for Incremental Shift Keying.
Doppler shift	Frequency shift caused by relative motion between a transmitter and a receiver. Common in satellite communications.
Downconverter	A device to convert a signal from a higher frequency to a lower one.
Downlink	Frequency used by a satellite to transmit to the user.
Driven element	The element in an antenna that is connected to the output of the transmitter.
DRM	Digital Radio Mondiale. A digital mode designed to compress signals into reduced bandwidth. Not to be confused with DMR, a digital voice mode.
DSP	Digital signal processing.
D-Star	Digital Smart Technologies for Amateur Radio. Primarily a digital voice and data mode marketed by Kenwood and ICOM.
DSW	Russian for goodbye in CW. (Do svidaniya.)
DTCS	Digital Tone Coded Squelch. Alternative to the CTCSS system.
DTMF	Dual-tone Multi Frequency. A system used to transmit and receive numeric information such as phone number, remote radio commands, etc.
Dual-band antenna	An antenna designed to be used on two amateur radio frequencies; typically, 2 meters and 70 cm.
Dualwatch	Feature on some receivers allowing them to monitor two frequencies at once.

Term	Definition
Dummy load	A device for testing transmitting equipment without radiating a signal.
Duplex	Operation mode in which the transmit and receive frequencies are different, allowing simultaneous reception and transmission. Repeaters operate in duplex mode.
Duplexer	A device that allows duplex (bi-directional) communications on a single path. In a repeater, it isolates the receiver from the transmitted signal while allowing the receiver and transmitter to share a common antenna.
Duty cycle	The percentage of time a transmitter is operating at full power during a single transmission. Some transmission modes have inherently higher duty cycles than others.
DX	"Distance." In ham radio, usually contacts with amateurs in foreign countries, but more generally, communication over long distances.
DX Maps	Web site that shows "real-time" data on propagation by showing QSO's plotted on a map.
DX Toolbox	A suite of propagation information related software for DX'ers.
DXCC	DX Century Club. An ARRL sponsored club in which one earns membership by showing proof of contacts with at least 100 countries.
DXLab Suite	A suite of propagation information and logging software for DX'ers.
DX-pedition	A trip to a foreign, often exotic land for the purpose of operating ham radios and making as many contacts as possible, giving other hams the opportunity to "score" contacts with countries with whom ham contacts are rarely available.
Dynamic range	The difference, expressed in dB, between the noise floor of a component and the "loudest" signal it can process without excessive distortion.
E	
E region	The middle layer of the ionosphere (or the lowest during night hours.)
Earth ground	An electrical connection to the Earth, typically accomplished with a ground stake or, in some locales, a ground plate.
Earth station	An amateur station located on Earth, or within 50 km of Earth's surface, engaged in communications with space stations or with other Earth stations using objects in space.
Echolink	A system that connects amateur radio operators to repeaters via cell phones.
EIRP	Effective Isotropic Radiated Power. The sum of transmitter output power and antenna gain expressed in dBi, minus transmission line losses.

Term	Definition
Electric field	A field of energy created when an electric current passes through a conductor. (See also, "magnetic field.")
Electron current	The flow of electricity from a negatively charged location to a positively charged location.
Elevation	In the context of directional antennas, a direction in the vertical plane. Most often shown on a graph of polar coordinates, with 0 degrees corresponding to the far right side of the circular graph and degrees plotted counter-clockwise around the circular
ELF	Extremely Low Frequency RF. Defined by the ITU as 3 Hz to 30 Hz.
Elmer	A mentor for newly licensed (or not so newly licensed) amateur radio operators.
E-M-E	Earth-Moon-Earth propagation. Commonly known as moonbounce.
Emergency	A situation in which there is immediate danger to life, limb, or property.
Emergency traffic	Messages containing information with life or death urgency, or dealing with requests for aid in an area experiencing an emergency.
EMF	Electromotive force; the force that pushes a current through a conductor.
EMI	Electro-magnetic interference. Often caused by battery chargers and DC to AC inverters.
Emission	A transmitted signal.
End-fed	Applied to antennas with a feed point at one end, as opposed to somewhere in the middle. Since only one-half of the transmitter's output is connected to the antenna, the other half must be connected to ground or to a ground plane.
ERP	Effective Radiated Power. The sum of the transmitter's output power and the gain of the antenna, minus transmission line losses. ERP is almost always a different value than PEP, usually higher. Most ham radio legal power limits are in watts PEP, but the 60 meter band power limit is in watts ERP, Estimated Radiated Power.
E-skip	Propagation by refraction from the E layer of the ionosphere.
EU	Europe.
Eyeball QSO	A face-to-face conversation with another ham.
F	
F connector	Type of connector often used in UHF installations, up to 1.2 GHz.

Term	Definition
F region	The highest layer of the ionosphere. During the day, the F region is subdivided into F1 and F2.
Fading	Signal reduction due to ionospheric effects.
Fan dipole	A multi-band center-fed wire antenna consisting of various lengths of wires strung between two points.
Farad	The unit of capacitance. A capacitor has 1 farad of capacitance if 1 coulomb of charge causes a potential difference of 1 volt.
FB	Fine Business. "Well done," or "Awesome, dude."
FCC	Federal Communications Commission.
Feed point	Point where the feed line connects electrically to the antenna.
Feedline	The cable or other conductor that connects the transmitter output to the antenna.
FET	Field Effect Transistor.
Field Day	An annual event, held over the fourth full weekend of June, in which hams worldwide set up temporary stations and operate "from the field." Individuals and ham radio clubs participate. Since part of the purpose of Field Day is to showcase our hobby for th
Filter	A circuit designed to pass only selected frequencies.
Fist	The sending style of a CW operator. Something like the operator's CW "accent" or "voice."
fldigi	Multiple digital mode software.
FM	Frequency modulation.
Form 605	The FCC form that is your application for a new or upgraded amateur radio license.
Foxhunting	see "bunny hunt."
Frequency	The number of times per second a signal passes through charge levels of zero, maximum positive, maximum negative, and back to 0.

Term	Definition
Frequency Bands (ITU)	Low Frequency: 30 kHz – 300 kHz Medium Frequency: 300 kHz – 3000 kHz (3 MHz) High Frequency: 3 MHz – 30 MHz Very High Frequency: 30 MHz – 300 MHz Ultra High Frequency: 300 MHz – 3000 MHz (3 GHz) Super High Frequency: 3 GHz – 30 GHz Extremely High Frequency: 30 GHz – 300 GHz Tremendously High Frequency: 300 GHz – 3,000 GHz
Frequency coordination	Local allocation of repeater input and output frequencies intended to reduce interference.
Frequency coordinator	A local group that recommends practices and implementations of frequency coordination.
Frequency discriminator	A type of detector used in FM receivers.
Frequency Modulation	A system of transmitting information by modulating the frequency of an RF carrier wave.
Frequency privileges	The set of frequencies on which an amateur is authorized to transmit, based on their class of license.
Front-end overload	A form of distortion caused by a strong signal overpowering the receiver's "front-end", its RF amplifier.
FRS	Family Radio Service. An unlicensed service that uses low-power radios operating in the UHF (460 MHz) band.
FSCW	A CW mode in which dots and dashes are transmitted at slightly different frequencies for better intelligibility.
FSK	Frequency Shift Keying.
FSK441	See "WSJT"
FSTV	Fast Scan TV.
FT8	See "WSJT"
Fuse	A device to protect an electrical circuit from excessive amperage.
Fusion	Yaesu's proprietary digital voice and data system.

G

Term	Definition
Gain	Generally, a positive difference in power, such as provided by an amplifier. Gain is expressed in dB. In the case of antennas, the ratio radiated power in the direction of most radiation versus either an isotropic antenna (dBi) or a 1/2 wave dipole antenn
Gel cell	A type of lead-acid battery that uses a gelled electrolyte, rather than liquid acid.
General coverage receiver	Usually an HF radio that will receive a wider range of frequencies than the amateur bands. Often, general coverage receivers cover frequencies from below the commercial AM broadcast band through at least the 10 meter amateur band.
General coverage receiver	A receiver capable of receiving a wide range of frequencies rather than specific bands of frequencies. Typically, general coverage receivers cover from below the AM broadcast bands through the amateur HF bands.
GFCI	Ground fault circuit interrupter. A device intended to prevent electrical shock by disconnecting a circuit when the current in the hot and neutral legs is not balanced.
GFI	See "GFCI"
GHz	Gigahertz. 1 GHz = 1,000,000,000 Hz
GMRS	General Mobile Radio Service. A licensed service operating in the UHF (460 MHz) range.
GOTA	Get On The Air. A station available for supervised public use at events such as Field Day.
Grace period (license renewal)	The time -- two years -- the FCC allows after a license has expired during which the holder of the license may renew without retesting. The license is not valid until renewed.
Gray line	A form of HF propagation available from north to south and vice versa as the Earth passes from day to night or vice versa.
Grid square	A location identifier, such as CN88xa.
Ground plane antenna	A 1/4 wavelength vertical antenna that employs horizontal radials extending from its base.
Ground radials	Elements of a vertical antenna that are horizontal relative to the ground (or close to horizontal.) Most vertical antennas require ground radials to work properly. Radials are not safety grounding and, in fact, are not electrically connected to the Earth.
Ground rod	A copper or copper-clad steel stake driven into the ground for the purpose of making an electrical connection with Earth.
Ground strap	Heavy-duty copper strap used for ground connections, particularly in lightning ground systems.

Term	Definition
Ground wave propagation	Propagation by radio waves that travel parallel to the surface of the Earth.
H	
HAAT	Height Above Average Terrain. For antenna and propagation calculations.
Half-duplex	See "semi-duplex"
Half-wave dipole	A basic antenna, usually consisting of two lengths of wire stretched horizontally with an insulator and the feed point in the center. The antenna is approximately one-half wavelength long relative to the desired operating frequency.
Ham	An amateur radio operator. The origin of the term is unknown, but most likely was a co-opting of a pejorative used by professional telegraphers near the time of the inception of amateur radio. Almost certainly not from "the initials of the original three amateurs" (or some variation on that theme.)
HamSphere	A virtual ham radio transceiver app for cell phones and computers. Somewhat analogous to Echolink, only for HF.
Hamvention	A gathering of amateur radio operators. Hamventions may be national, regional, or even local, and usually include vendors of amateur radio gear, speakers on various ham related topics, a swap meet, a closing dinner with a "star" speaker, and lots of camar
Harmonic	A multiple of a fundamental frequency. Colloquially, a ham's child.
Health and welfare traffic	Messages regarding the well-being of people in a disaster area. Health and welfare traffic has lower priority than "emergency" and "priority" traffic.
Heat sink	The heavy fins placed on top of a high heat component such as the final drive transistors in a transceiver. Designed to prevent excessive heat build-up.
Hellschreiber	An image transmission mode, similar to faxing.
Henry	The unit of inductance. Equal to an electromotive force of 1 volt in a closed circuit with a uniform rate of change of current of 1 ampere per second.
Hertz	See "Hz"
Hex Beam Antenna	A directional HF antenna in the shape of a horizontal hexagon. Typically multi-band, and intended for installations where low visual impact is desired.
HF	High Frequency RF. Defined by the ITU as 3 MHz to 30 MHz
HI HI	Morse code equivalent of laughing.

Term	Definition
High pass filter	A circuit or device that allows all frequencies above a defined frequency to pass while blocking all frequencies below that defined frequency.
Homebrew	Some piece of equipment that is homemade. As a verb, the act of making equipment at home.
Hotspot	A link between a digital voice radio (DRM, Fusion, D-Star, etc.) and the internet. Allows the operator to connect worldwide with a VHF/UHF transceiver without using a wide coverage repeater.
HT	A handheld transceiver or "handy talkie."
Hz	Hertz. The measurement of frequency. 1 Hz = 1 cycle per second.
I	
Iambic	A type of Morse code key with two paddles. Pressing one paddle sends a dash, holding it sends a series of dashes. Pressing the other sends a dot, and holding it sends a series of dots. Pressing both paddles simultaneously sends dot dash. Skilled operators can achieve speed gains with this system vs. a "straight key."
IARU	International Amateur Radio Union. The association of national ham organizations, such as the ARRL.
IC	An Integrated Circuit. A "chip."
ICOM	Amateur radio equipment manufacturer. One of the "Big Three."
IF	Intermediate Frequency. Part of the superheterodyne system of radio reception and detection.
IF shift	A circuit that shifts the IF frequency from a center frequency to help reduce interference.
IFK	Incremental Shift Keying. A digital mode.
Image frequency	A frequency offset from the desired signal by double the receiver's Intermediate Frequency. Related to image rejection.
Image rejection	A receiver specification that shows how well the receiver rejects interference from stations on frequencies that are double the receiver's Intermediate Frequency.
IMD	Inter-modulation distortion.
Impedance	The opposition to the flow of current in a circuit. Impedance consists of resistance and capacitive and inductive reactance.

Term	Definition
Impedance matching network	See "Antenna tuner"
Inductance	The amount of magnetic storage available in an inductor. Measured in Henrys.
Inductor	A component -- typically a coil -- that stores energy in a magnetic field.
Input frequency	With regard to repeaters, the frequency the repeater is "listening to." To use the repeater you must transmit on the repeater's input frequency.
Insulator	Any material or component that blocks the flow of electrical current.
Inverter	A device that converts DC to AC.
Ionizing radiation	Electromagnetic radiation that has sufficient energy to ionize atoms, producing negative and positive ions. Ultraviolet, X-rays, and gamma rays are ionizing radiation. RF is not.
Ionosphere	A region of ionized gasses above the stratosphere.
IOTA	Islands on The Air. A ham radio activity based on operating from as many islands as possible.
IRLP	The Internet Radio Linking Project that allows amateur operators to join a global network of conversations.
ISCAT	See "WSJT"
Isotropic antenna	An imaginary "ideal" antenna that radiates perfectly equally in all directions with zero loss. Used as a reference point to compare the gain of a directional antenna. Directional gain can be specified in "dBi", meaning "dB relative to an isotropic antenna," or in "dBd", meaning "dB relative to a half-wave dipole antenna."
ISS	International Space Station.
ITU	The International Telecommunications Union. An agency of the United Nations that is responsible for issues that concern information and communication technologies.
ITU Regions	The ITU divides the world into 3 regions. Region 1 comprises Europe, Africa, the former Soviet Union, Mongolia, and the Middle East west of the Persian Gulf, including Iraq. Region 2 covers the Americas including Greenland, and some of the eastern Pacific
	J
JOTA	Jamboree On The Air. An annual Boy Scout amateur radio event.

Term	Definition
J-pole	A type of omnidirectional half-wave antenna, usually for 2 meters and 70 cm, that resembles the shape of a letter J.
JT4	See "WSJT"
JT65	See "WSJT"
JT9	See "WSJT"
JTMS	See "WSJT"
Jumper	A short piece of wire or cable used to connect two parts of a circuit or two pieces of equipment.
K	
Kennelly-Heaviside Layer	Alternate name for the E layer of the ionosphere.
Kenwood	Amateur radio equipment manufacturer. One of the "Big Three."
Kerchunk (a repeater)	To press the PTT button momentarily to test whether one is hitting a repeater. A repeater that is activated will leave a tell-tale "squelch tail" that can be heard after the PTT is released. Kerchunking is bad form and illegal -- it constitutes an unident
Key	Hand-operated device used to produce Morse code.
Keyer	Electronic device for sending Morse code semi-automatically.
kHz	Kilohertz. 1 kHz = 1,000 Hz
Knife-edge diffraction	The bending of a signal by tall buildings and mountains.
L	
LCD	Liquid Crystal Display.
LED	Light Emitting Diode.
Lid	Amateur radio slang for an unskilled operator.
Lightning ground	In amateur radio, a system designed to divert lightning strikes to the antenna or tower to ground, rather than into equipment or a residence.

Term	Definition
Li-Ion	A Lithium Ion battery.
Limiter	A circuit in a receiver that reduces variations in audio levels.
Line-of-sight propagation	Propagation in a straight line from one station to another.
Lobe	An area in the radiation pattern of an antenna.
Logbook of The World	See, "LoTW."
LOS	Loss of Signal from a satellite.
LoTW	"Logbook of The World." An online logging and QSO confirmation system run by the ARRL.
Low pass filter	A circuit or device that allows all frequencies below a defined frequency to pass while blocking all frequencies above that defined frequency.
Lowfer	An amateur radio operator who operates on the low frequency 2,200 meter band.
LPDA	Log Periodic Dipole Array. A type of multi-band antenna.
LSB	Lower sideband modulation. In LSB, the upper sideband and the main carrier are "suppressed" - filtered out -- leaving only the lower sideband. A form of amplitude modulation.
LUF	The Lowest Useable Frequency; the lowest frequency radio signal that will reach a given destination via ionospheric propagation.
M	
Machine, The	Colloquial for a repeater.
Magnetic field	A field of energy created when an electric current passes through a conductor. (See also, "electric field."
Maidenhead Grid System	A ham-created grid system for locating any spot on Earth within a "grid square." The system encodes latitude and longitude in a series of letters and numbers, such as CN88bh. The system grew out of an amateur radio meeting held at Maidenhead, England; hence, the name.
Malicious interference	Deliberate disruption of legal radio transmissions.
Maritime mobile	Amateur radio operations from aboard a marine vessel.

Term	Definition
MARS	Military Affiliate Radio Service.
Matchbox	See "antenna tuner"
Maximum Permissible Exposure	The maximum amount of RF radiation to which a human being may legally be exposed.
Mayday	International distress symbol on phone. From French, "m'aidez" -- "help me."
MCW	Modulated Continuous Wave. A method of sending Morse code using audio tones, rather than with momentary pulses of carrier. Commonly used by repeaters for station identification.
Meteor scatter	Propagation from the trails of ions left in the ionosphere by falling meteors.
MF	Medium Frequency RF. Defined by the ITU as 300 kHz to 3,000 kHz (3 MHz.)
MFSK	Multiple Frequency Shift Keying.
MHz	Megahertz. 1 MHz = 1,000,000 Hz
Microphone	A device to convert sound energy into electrical energy.
MMSSTV	Slow-scan TV software.
MMTTY	Multiple digital HF mode software.
Mobile	In amateur radio, generally a transceiver mounted in a vehicle.
Mode	In the context of transmission and reception, the form of modulation used to transmit information. For instance, CW, SSB, Phase Shift Keying, etc.
Modem	A "modulator/demodulator." Converts data into a radio signal and vice versa.
Modulation	The application of information to a carrier signal by systematically varying one or more dimensions of that carrier, such as amplitude, frequency, or phase angle.
Modulation Index	See "deviation."
Modulator	A circuit in a transmitter that alters a dimension (or dimensions) of a carrier wave in order to transmit information.
Monitor mode	In packet radio, a mode in which everything on a packet frequency is displayed without regard to whether the messages were addressed to the monitoring station or not.

Term	Definition
MOSFET	Metal Oxide Semiconductor Field Effect Transistor.
MP73N	A narrow SSTV mode.
MSK144	See "WSJT"
MT63	A weak signal digital mode used in MARS net traffic.
MUF	The Maximum Useable Frequency; the highest frequency radio signal that will reach a given destination via ionospheric propagation.
Multimeter	An instrument for measuring voltage, current, and resistance. Many modern multimeters measure more values as well.
Multimode transceiver	A transceiver capable of sending and receiving multiple modes, such as CW, SSB, AM, and FM.
MultiPSK	Multiple digital HF mode software.
N	
National Electrical Code (NEC)	A set of safety guidelines for any electrical installation including towers, grounding systems, power systems, and antennas. Not all locale's local electrical code is an exact match to the NEC.
National Traffic System	An on-air network with the purpose of passing "traffic" - messages that may be between hams or between members of the public.
NB	see "Noise blanker."
NBFM	Narrow Band FM. Frequency Modulation with a deviation ratio of less than 0.5; 0.2 is common.
NCDXF	The Northern California DX Foundation, a private foundation with the purpose of supporting amateur radio and scientific projects with funding and equipment. In 1979 the NCDXF Board launched the NCDXF/IARU International Beacon Project to provide a mechanis
NCS	See "Net control station"
NCVEC	National Conference of Volunteer Examiner Coordinators. The national association of VEC's, and official keepers of the standard question banks for license exams.
Negative feedback	The process of feeding back a certain amount of the transmitter's output to the input of the final drive 180 degrees out of phase to prevent the amplifier from destroying itself.
Net control station	The operator in charge of conducting an on-air network.

Term	Definition
Ni-CD	A Nickel Cadmium battery.
NI-MH	A Nickel Metal Hydride battery.
Noise	Unwanted electromagnetic energy.
Noise blanker	A circuit in a receiver that blocks intermittent noise pulses.
Noise Figure	A receiver specification denoting the difference, in decibels, between the noise output of the receiver and an "ideal" receiver with the same gain and bandwidth. Lower numbers are better.
Noise floor	The lowest level signal a receiver can "hear." Any lower level signal is below the noise floor and therefore undetectable.
Notch filter	A very narrow bandstop filter.
NPN transistor	A transistor consisting of a layer of P-type material between two layers of N-type material.
NR	Noise Reduction.
NTS	See "National Traffic System."
Null	A position in the pattern of an antenna where its reception or transmission is at minimum.
NVIS	Near Vertical Incidence Skywave. Propagation created by a high angle radiation of signal -- the closer to straight up, the better for NVIS. For short to medium distance communication via HF and, more rarely, VHF.
NXDN	Digital voice and data mode used by Kenwood and ICOM.
O	
OCF	Off-Center Fed dipole antenna. A type of multi-band dipole.
Off-center fed	Applied to antennas with a feed point that is neither at the end nor in the center. Off-center fed dipoles can be multi-band antennas.
Offset	See "repeater offset"
Ohm (Ω)	The unit of electrical resistance. 1 ohm = a resistance that will pass a current of 1 ampere when subjected to a potential difference of 1 volt. Also the unit of impedance and of reactance.
Ohmmeter	An instrument used to measure resistance.

Term	Definition
Ohm's Law	A fundamental law of electronics that describes the proportional relationship among voltage, current, and resistance.
OM	"Old Man." Term of endearment for a male ham.
One-way communications	Communications that are not intended to be answered.
Op-amp	Short for Operational Amplifier. A high-gain solid state amplifier widely used in many applications.
Open circuit	An electrical circuit with a break in the circuit that prevents the current from completing the path to ground. An open circuit might be purposefully created with a switch, or accidentally created by a faulty connection.
Open repeater	A repeater that may be used by all hams possessing the proper license for its frequency and mode of operation.
OSCAR	Orbital Satellite Carrying Amateur Radio. Name of a series of ham radio satellites.
Oscillator	A circuit that creates a signal of a particular frequency.
Oscilloscope	An instrument used to display and measure various dimensions of a signal, most often the waveform.
Output frequency	With regard to repeaters, the frequency on which the repeater is "talking." To use the repeater, you must listen to the repeater's output frequency.
P	
P25	A digital voice mode. More formally, APCO P25.
P5	Call sign prefix of North Korea; presently the single most difficult stations with which to make a contact, although Yemen, 70, is no picnic either.
PA	Power amplifier
Packet radio	A mode of digital communication in which information is broken down into small "packets" of information for transmission, then reassembled at the receiver end.
PACTOR	Digital mode used mostly on the HF bands for text messaging.
Parallel circuit	An electrical circuit in which there are two or more paths.
Parasitic beam antenna	see "beam antenna"
Term	Definition

Term	Definition
Parasitic element	A passive element of an antenna.
Part 97	The body of FCC rules and regulations that create and regulate the Amateur Service.
Passband	The range of frequencies passed by a filter.
Peak envelope power (PEP)	The average power output of a transmitter at the highest amplitude.
PEI	Peak envelope current. One half of the equation for peak envelope power.
PEP	See "peak envelope power."
Perigee	The point in an orbiting object's orbit where it is closest to the object around which is it orbiting.
Period	The time in seconds (or fractions of seconds) for a complete wave to pass a stationary point. The reciprocal of frequency.
PEV	Peak envelope voltage. One half of the equation for peak envelope power.
Phase Modulation	A system of transmitting information by modulating the phase of an RF carrier wave.
Phone	Communication by voice.
Phone emission	FCC term for voice or other sound transmission.
Phone patch	See "autopatch."
Photovoltaic	Shorthand for a photovoltaic cell, i.e., a solar cell.
PL	See "CTCSS"
PL-259	See "UHF connector"
PM	Pulse modulation or phase modulation.
PNP transistor	A transistor consisting of a layer of N-type material between two layers of P-type material.
Polar coordinates	A graphing system that defines a point by its distance from the center of a graph (the "pole") and the angle to the point from the pole.

Term	Definition
Polarization	The orientation, relative to the Earth, of an electromagnetic wave's electric field. A wave with both vertical and horizontal polarization is said to be circularly polarized.
Portable device	A transmitter designed to be easily operated while being carried and, for FCC purposes, with an antenna intended to be within approximately 20 cm of a human body.
Power supply	A circuit or device that supplies power for electronic equipment. Power supplies must supply adequate amperage at the correct voltage.
Priority traffic	Emergency related traffic that is not as urgent or important as Emergency traffic.
Priority watch	A feature of some receivers. The priority frequency is checked for traffic periodically, no matter where the VFO is set.
Product detector	A circuit in a receiver that detects SSB and CW signals.
Propagation	The process by which a radio wave is carried from a transmitter to a receiver. Informally, "everything that happens to our signal after it leaves our transmitting antenna."
PropLab Pro	High-end propagation calculation software.
Prosigns (Procedural signals)	Over-the-air shortcuts, such as "CQ", "73", etc. Originally used by telegraphers to speed message transmission, they are still in wide use in the amateur community.
PSK31	A type of radio-teletype using Phase Shift Keying. Very narrow bandwidth of 31 Hz.
PTT	Push-To-Talk
PWR	Power
Q	Describes the response of a resonant circuit over a specific bandwidth.
Q	
Q signals	Three-letter symbols that begin with Q and are used on CW (and sometimes on phone) to save time and increase accuracy of communication.
QCWA	Quarter-Century Wireless Association. A club for amateurs who have held a license for 25 years or longer.
QPSK	Quadrature Phase Shift Keying. A digital text and data mode.
QRA64	See "WSJT"

Term	Definition
QRP	Low power operation, usually 5 watts or less.
QRPP	Extremely low power operation, usually less than 1 watt.
QRSS	Extremely slow-speed CW transmissions, usually less than one character per minute.
QRZ	"Who is calling me?" Also the name of a popular ham radio web site, qrz.com, which offers easy look-ups of call signs, grid squares, etc.
QSL Bureau	An organization that provides a collection and distribution point for QSL cards, for the purpose of saving everyone's postage costs.
QSL card	A paper confirmation, almost always a postcard, of a contact with another station.
QSO	A contact.
QSO Party	Ham radio contest.
QST Magazine	One of the two major general interest amateur radio magazines, the other being CQ Magazine.
Quad antenna	A type of directional antenna constructed of elements in the shape of squares.
Quad-band antenna	An antenna designed to be used on four amateur radio frequencies.
R	
RACES	The Radio Amateur Civil Emergency Service.
Radials	See "ground radials"
Radio horizon	The farthest point a VHF or higher frequency signal can travel under normal conditions. Slightly beyond the visual horizon.
Rain scatter	Propagation off a (dense) rain front.
Random wire antenna	An antenna consisting of an end-fed random length of wire, usually longer than one wavelength of the desired band of operation. An antenna tuner is almost always required for operation, as is a solid Earth ground. According to Arizona ham Patrick Lambert
Receiver	A device that detects, amplifies, and retrieves the information from a signal.

Term	Definition
Rectangular coordinates	A graphing system that defines a point by its horizontal and vertical distance from the intersection of two axes, one of which is horizontal and the other vertical. Also known as "Cartesian coordinates." Less formally, it's the "X, Y" system of graphing.
Reflected power	A product of SWR. Non-radiated power that is dissipated as heat.
Reflector element	A passive antenna element, typically located behind the driven element.
Regulator	A circuit or component that maintains a constant voltage. Part of a power supply.
Repeater	A system consisting of a receiver, a transmitter, and some means of controlling those units such that they receive signals and re-transmit them on a slightly different frequency for the purpose of extending the range of transmission.
Repeater offset	The frequency difference between a repeater's input frequency and its output frequency.
Resistance	The opposition posed by a circuit or component to the flow of electricity.
Resistive load	An electrical load that presents only resistance, with no capacitive or inductive reactance. An ideal antenna, for example, presents a purely resistive load that has the same resistance as the impedance of whatever is feeding it.
Resistor	Any material or component that opposes the flow of electrical current.
Resonant	A circuit, antenna, or even a mechanical system is said to be resonant when a relatively low amplitude, periodic stimulus of the same period as the natural vibration period of the antenna, circuit, or system produces a vibration of a large amplitude. A ci
Resonant circuit	A circuit designed to resonate at a particular frequency, typically consisting of a capacitor, an inductor and a resistor. Multiple uses.
Rettysnitch	Companion device to the Wouff Hong.
Reverse (function)	A feature of some VHF/UHF transceivers to allow quickly switching from the output frequency of a repeater to the input frequency, for purposes of enhancing reception.
RF	Radio Frequency. Electromagnetic waves with frequencies between 3 Hz and 3,000 GHz.
RF burn	A skin burn caused by contact with exposed RF voltages. A symptom of poor RF grounding.
RF ground	A grounding system designed to divert stray RF energy to ground.
RFI	Radio Frequency Interference. Disfunction in a piece of electronic equipment created by RF.

Term	Definition
Rig	Amateur radio term for a transmitter, receiver, transceiver, or radio transmission system.
Ripple	In the context of a DC power supply, a waveform remaining on the DC after it is filtered.
RIT	Receiver Incremental Tuning. Allows fine tuning of the receiver frequency for better SSB reception without changing the transmit frequency.
RMS	Root Mean Square. Relates to the mathematical concept of root mean square and applies to AC voltage. Put simply, the RMS voltage is the "DC equivalent" voltage; in other words, an AC voltage of 120 volts RMS will produce the same amount of power as a DC
ROS Digital	Digital text mode for low signal conditions. "Ros" is the inventor's last name.
RSGB	Radio Society of Great Britain
RSQ code	Readability, Signal (strength), Quality. A signal report code similar to the RST code for amateur digital transmissions.
RST	A code of three numbers used to indicate quality of reception of a signal. R is for readability, S is for signal strength, and T is for tone and applies only to CW transmissions.
RSV code	Readability, Signal (strength), Video. A signal report code similar to the RST code for amateur television.
RTTY	Radio-teletype. A low bandwidth mode of transmitting text messages, originally to a teleprinter but now to a computer.
Rubber duck/rubber ducky	A flexible, shortened antenna used mostly on handheld transceivers. Usually covered with black plastic. Legend says they were named by a very young Caroline Kennedy when the Secret Service got the then-new antennas for their radios.
RX	Receive or receiver.
S	
S meter	A signal strength meter.
S units	Graduations on a signal strength meter.
S/N	Signal-to-noise ratio.
Safety ground	A system designed to divert high voltages on equipment cases and other items exposed to human contact directly to ground, rather than through the human.

Term	Definition
Safety interlock	A mechanically activated switch that switches off AC power to a piece of equipment when the equipment is opened.
Scan	A feature in some receivers that continually sweeps through a range of frequencies or a set of frequencies searching for signals.
SDR	Software defined radio. A form of receiver or transceiver that digitizes the incoming RF so a computer can decode the signals.
Selectivity	The ability of a receiver to discriminate between two closely spaced signals.
Semiconductor	A substance which is normally an insulator but which can transform to a conductor under certain physical and electrical circumstances.
Semi-duplex	An operation mode in which transmit and receive functions are on different frequencies alternatively.
Sensitivity	The ability of a receiver to detect weak signals.
Series circuit	An electrical circuit in which there is only one path.
Series-parallel	An electrical circuit that combines parallel and series elements.
SFI	Solar Flux Index
Shack	An amateur radio station location and associated equipment.
SHF	Super High Frequency RF. Defined by the ITU as 3 GHz to 30 GHz.
Short circuit	An electrical circuit in which a fault of some kind is taking the current to ground instead of through the desired path.
Signal reports	See "RST"
SignalLink	A brand of device for connecting a computer to an amateur transceiver via audio. The device substitutes for the computer's sound card and provides PTT control when connected with the proper cable.
Simplex	Operation mode in which the transmit and receive frequencies are the same.
SK	1. "Ceasing transmissions now, not expecting a reply." 2. A deceased amateur radio operator.
SKED	A pre-arranged contact. "I have a sked with AF7KB at 2200 Zulu."

Term	Definition
Skip zone	An area between the end of a station's ground wave coverage and the beginning of its skywave coverage.
Skywarn	A network of trained volunteer amateur radio operators who serve as storm spotters.
Skywave propagation	Propagation via ionospheric refraction.
SLF	Super Low Frequency RF. Defined by the ITU as 30 Hz to 300 Hz.
SMA	Sub-Miniature. A type of connector, often used in VHF/UHF handheld transceivers.
SOS	International distress symbol in Morse code.
SOTA	Summits on The Air. A ham radio activity based on operating from as many mountain summits as possible.
SP	Speaker
Space station	In amateur radio, an amateur station more than 50 km above Earth's surface.
Specific absorption rate	The rate at which RF energy is absorbed into a human body.
Spectrum analyzer	An instrument used to display and measure the amplitude of signals present in a particular range of frequencies.
Splatter	Interference to stations on nearby frequencies. Caused by overmodulation.
Split mode	An operating mode in which the transmit and receive frequencies are different from each other.
Split operation	Operating with transmit set on one frequency and receive on another. Often used by DX-peditions because of the enormous number of incoming transmissions they must field and answer.
Sporadic E	Semi-random propagation from anomalous regions of high free electron density in the E region.
Spread spectrum	A modulation and transmission system that spreads a signal over a wide bandwidth.
SQL	see "squelch"
Squelch	A function that mutes audio output for certain conditions, primarily in the absence of a detectable carrier.
SS	Spread spectrum modulation.

Term	Definition
SSB	Single sideband modulation. A form of amplitude modulation in which one sideband and part or all of the main carrier are removed from the transmitted signal.
SSN	Smoothed Sunspot Number
SSTV	Slow Scan Television. A mode of transmitting still pictures via amateur radio.
Standing wave	In physics language, the vector sum of two waves. For amateur radio, usually SWR.
Station log	A record of contacts made, signal reports, etc. Often computerized.
Straight key	One of the simplest devices for sending Morse code. See any Western movie with a scene in a telegraph office.
Sunspot cycle	The approximately 11 year cycle of average sunspot number waxing and waning.
Sunspots	Spots on the surface of the Sun that appear darker than the surrounding material. High numbers of sunspots correlate to improved HF propagation conditions.
Superheterodyne receiver	A receiver in which the incoming RF is down-converted to an intermediate frequency. The rest of the RF stages in the radio are optimized to operate at that frequency.
Susceptance	The reciprocal of reactance. Measured in siemens.
SWR	Standing wave ratio. The ratio of the maximum and minimum voltages present on a transmission line. Also the ratio of any difference in impedance between the antenna and the transmission line, assuming the antenna is a purely resistive load.
SWR meter	An instrument for measuring SWR in a transmission system.
Symbol	In the context of baud rate a symbol is a signal change that indicates information. For the simplest example, consider the Morse code for the letter A, dit dah. It consists of four symbols. The dit is a symbol, the space between the dit and the dah is another symbol, the dah is a third symbol, and the (implied) space after the dah is a fourth symbol.
System Fusion	See "fusion."
T	
Tactical call signs	"Call signs" used to identify particular functions or locations; typically during special events or emergencies. They do not replace standard call signs for station identification purposes.
Talkaround	Simplex operation.

Term	Definition
Tank circuit	See "resonant circuit."
Telegraphy	Text-based modes such as CW, or RTTY. As opposed to phone.
Temperature inversion	A weather condition in which a layer of warm air rests on top of a layer of cooler air. Can lead to tropospheric ducting propagation.
Terminal	A piece of equipment that can replace a computer in a packet radio system.
Third-order Intercept	A receiver specification denoting how subject a receiver is to intermodulation distortion. Higher dB figures are better.
Third-party communications	Messages passed from one amateur radio operator to another on behalf of a third, unlicensed, party.
Third-party communications agreement	An agreement between the US and another country allowing amateurs to pass third-party communications between the countries.
THROB	A digital mode based on tone pairs.
Ticket	Common name for an amateur radio operator/primary station license.
Time Out Timer	See "TOT"
Time-out	A repeater feature that temporarily shuts down the transmitter portion of the repeater after a preset amount of transmission time. This prevents overheating of the transmitter and limits how long one person may talk on a repeater.
TNC	Terminal Node Connector modem for data communications. Also, a type of antenna connector.
Topband	Affectionate nickname for 160-meter band.
TOR	Teleprinting Over Radio. There are a number of TOR modes.
Toroid	In the context of inductors, an inductor with a doughnut shaped core, whether air or some other substance.
TOT	Time-out timer. A time limit function for repeater transmitters to prevent them from overheating from long periods of uninterrupted transmission.
Transceiver	A combination unit capable of transmitting and receiving.
Transient	A brief spike of energy, especially in regards to power line supplied voltage.

Term	Definition
Transistor	Any of a wide array of solid-state devices with multiple uses including amplification.
Transmatch	See "Antenna tuner"
Transmission line	See "feedline."
Transmitter	A device for generating RF signals.
Transverter	A device for converting received and transmitted frequencies either up or down. Might be used with a 10-meter radio to operate on 6 meters, with the received signal downconverted and the transmitted signal upconverted.
Trapped dipole	Multi-band dipole antenna. It is constructed with low-pass filters in the elements to create the electrical equivalent of several different lengths of antenna.
Troposphere	The lowest region of the Earth's atmosphere.
Tropospheric bending	A form of propagation typically created by mild temperature inversions.
Tropospheric ducting	A type of propagation that can occur during temperature inversions. Typically a VHF/UHF phenomenon.
TrueTTY	Multiple digital HF mode software.
TU	Thank you in CW.
Tuned circuit	See "resonant circuit."
Tuning Step	The increment that a transceiver changes frequency for one "step" of the tuning control.
TVI (Television interference)	Disruption of television reception caused by RF.
TX	Transmit or transmitter
TXCO	A Temperature Compensated Crystal Oscillator, a type of oscillator with excellent frequency stability.
U	
UFB	Ultra Fine Business. "Magnificent, splendid!" "Dude -- you crushed it."
UHF	Ultra High Frequency RF. Defined by the ITU as 300 MHz to 3,000 MHZ (30 GHz)

Term	Definition
UHF connector	Also known as a PL-259 plug, for coaxial cable.
ULF	Ultra-Low Frequency RF. Defined by the ITU as 300 Hz to 3,000 Hz (3 kHz.)
Unbalanced line	A feed line with one conductor at ground potential, such as coaxial cable.
Unun	A device for connecting an unbalanced source to an unbalanced load.
Uplink	Frequency used by a user to transmit to a satellite.
USB	Upper sideband modulation. In USB, the lower sideband and the main carrier are "suppressed" - filtered out -- leaving only the upper sideband. A form of amplitude modulation.
UTC	Universal Time Coordinated. While it is, for most practical purposes, the same as Greenwich Meridian Time, it is a microscopically different time standard. Most ham radio related times-of-day are given in UTC.
	V
VA	Volt Amperes. The measure of Apparent Power (as opposed to Real Power.)
VAC	Volts of Alternating Current.
Vanity Call Sign	A call sign issued by the FCC upon a request from an operator. Certain rules and fees apply. https://www.fcc.gov/amateur-call-sign-systems
Varactor	A specialized diode who capacitance varies as the bias voltage is varied.
VCO	A Voltage Controlled Oscillator; A VCO's frequency of oscillation is controlled by an applied voltage. Typically found in phase locked loop circuits.
VE	A Volunteer Examiner; someone who administers ham radio license exams.
VEC	A Volunteer Examiner Coordinator; an organization under whose umbrella Volunteer Examiners administer ham radio license exams.
Vertical antenna	An antenna in which the radiating element is vertical relative to the ground.
VFB	Very Fine Business. "Extraordinarily well done," or "TOTALLY awesome, dude!"
VFO	Variable Frequency Oscillator. An oscillator in a receiver or transmitter that provides an adjustable frequency. Colloquially, the tuning knob.

Term	Definition
VHF	Very High Frequency RF. Defined by the ITU as 30 MHz to 300 MHz
VIS	Vertical Interval Signaling. Information in an SSTV transmission that conveys the mode of transmission in use.
Visual horizon	The farthest point one can see by line-of-sight.
VLF	Very Low Frequency RF. Defined by the ITU as 3 kHz to 30 kHz.
VM	See "Volunteer Monitor."
Volt (V)	The unit of electromotive force. 1 volt = the difference of electric potential that would drive 1 ampere of current through 1 ohm of resistance.
Voltage	Electromotive force. The amount of charge in one location relative to another location. Informally, "electrical pressure." Measured in volts.
Voltmeter	An instrument used to measure voltage.
Volunteer Monitor	A member of the Volunteer Monitoring Program. A volunteer amateur operator who monitors the airwaves for FCC rules violations.
VOX	Voice Operated Transmission. Allows the presence of sound at a microphone to operate the PTT.
VSWR	See "SWR."
VUCC	"VHF/UHF Century Club." An ARRL award for verified contacts with a minimum number of Maidenhead grid locators per band.
W	
WAC	Worked All Continents. An award issued by the IARU to those who prove they have had a contact with at least one ham on each continent.
WAN	Worked All Neighbors. A station that generates many complaints about interference with neighborhood telephones and televisions. (Don't get this award.)
WARC	World Administrative Radio Conference.
WARC bands	Created by the WARC in 1979, they are the 30 meter, 17 meter, and 12 meter bands.
WAS	Worked All States. An ARRL award to those who prove they have made a contact with at least one station in each U.S. State.

Term	Definition
Watt (W)	The unit of measurement of the use of electrical power. 1 watt = a current of 1 amp at a voltage of 1 volt.
Wattmeter	An instrument for measuring power; in the case of amateur radio, this is usually the output of the transmitter. Wattmeters can be directional, measuring either forward or reflected power.
Waveguide	A carrier of microwaves to and from a radio to an antenna. A waveguide is a structure that guides electromagnetic waves (or, in other applications, sound waves) with little loss of energy by restricting the expansion of that wave. Also, a form of propagat
Wavelength	The physical length of one cycle of a signal. In any given medium, all radio waves travel at the same speed, so lower frequencies have longer wavelengths. To determine the approximate wavelength in meters of any frequency, divide the frequency, expressed
WAZ	An award from CQ Magazine recognizing amateurs who have proof of contact with a particular number of the "CQ DX Zones." The number varies by band and mode. See http://www.cq-amateur-radio.com/cq_awards/cq_waz_awards/cq_waz_award_types.html
WFM	Wideband FM
Wires-X	"Wide-coverage Internet Repeater Enhancement System." An internet linking feature of the Yaesu System Fusion digital voice and data system.
Wouff Hong	A torture device of unspeakable horror used for punishing hams who fail to follow common courtesy on the air. Many ham conventions feature a secretive late-night ceremony in which they are inducted into the mysterious Royal Order of the Wouff Hong.
WPX	"Worked Prefixes." An award from CQ Magazine recognizing amateurs who have proof of contact with a particular number of stations whose call signs begin with the many international call sign prefixes around the world.
WSJT	The Weak Signal Joe Taylor suite of weak signal digital transmission modes. Various modes can be used for meteor scatter, moonbounce, and other weak signal situations, including general purpose communication in poor propagation conditions. The software suite is available free at https://physics.princeton.edu/pulsar/k1jt/wsjtx.html
WSPR	Weak Signal Propagation Reporter. Part of the WSJT suite.
WSPRNet	A reporting tool for the WSPR system.
WWV	A radio station located in Fort Collins, CO, and operated by the National Bureau of Standards. It broadcasts the time of day, propagation reports and other information, and serves as a frequency standard.
WWVH	A station similar to WWV, broadcasting from Hawaii.
WX	Weather.

Term	Definition
X	
XCVR	Transceiver.
XIT	Transmitter Incremental Tuning. A feature on some HF radios that allows fine tuning of the transmit frequency without changing the receive frequency.
XTAL	Crystal.
XYL	An "ex-YL." Vintage ham term for "wife." "YL" is far more commonly used now, for reasons that are probably obvious.
Y	
Yaesu	Amateur radio equipment manufacturer. One of the "Big Three." Usually pronounced YAY-zoo.
Yagi antenna	A type of directional antenna with a single dipole radiating element, a reflector, and at least one director element, all mounted parallel to each other.
YF	Short for "wife."
YL	"Young lady." Might describe any female, but especially a female ham, or a ham's wife.
Z	
Zed	The letter "Z."
Zener diode	A specialized diode used for power supply regulation.
Zero beat	To tune precisely to a received frequency.
Zulu	In the context of time-of-day, UTC

Index

1.25 Meters	22
10 meter repeater offset	19
12 Meters	19
122.25 -123.00 GHz	32
13 Centimeters	26
134-141 GHz	32
15 Meters	18
160 Meters	13
17 Meters	18
2 Meters	21
2200 Meters	12
23 Centimeters	25
24.0-24.25 GHz	32
241-250 GHz	32
3 Centimeters	31
30 Meters	17
300-baud packet	15
304A	116
33 Centimeters	24
40 Meters	16
47.0-47.2 GHz	32
5 Centimeters	30
6 meter repeater offsets	20
6 Meters	20
60 Meters	15
630 Meters	13
70 Centimeters	23
76.0-81.0 GHz	32
80 Meters	14
9 Centimeters	27
A Ham Radio Auction	161
A1A	95
A2A	95
A3E	95
Aatest	159
ABR Industries	161
AC Log by N3FJP	159
Accrediting VEs	154
ACOM International	161
Administering VE requirements	152
AE	132
AG	132
A-index	115
Aircraft VHF Frequencies	36
Airlink Express	159
Alinco Electronics	167
Alltronics	161
Alpha Antenna	161
Alpha Delta Communications	161
Aluma Towers	161
amateur bands above 10.5 GHz	32
Amateur operator	119
Amateur Radio News	158
Amateur Radio Newsline	158
Amateur Radio on the International Space Station	165
Amateur Radio Organizations	158
Amateur Radio Repairs	162
Amateur Radio Service Center	161
Amateur radio services	119
Amateur Radio Software	159
Amateur station	119
American Radio Supply	161
Ameritron	161
Amidon Corp	161
Amidon Inductor Cores	161
Ampacity	96
AMSAT	159
AMTOR	148
Antenna analyzers	164
Antenna Efficiency	114
Antenna guying	164
Antenna Length, ¼ Wave Antenna	112
Antenna Length, ½ Wave Antenna	112
Antenna Lengths	103
Antennas US	161
Antique Electronic Supply	161
AOR, Inc.	161
Application for a modified or renewed license grant	126
Application for a vanity call sign	125
Application for new license grant	124
Applied Interconnect	164
Arcom Controllers	161
Arecibo Observatory	133
ARISS	165
Arlan Communications	161
Armed Forces Day Communications Test	129
Array Solutions, Inc.	161
ARRL	12, 21, 22, 90, 91, 158, 160
ASA, Inc. (Waterproof log books)	161
ASCII	148
Associated Radio	160
Authorized emission types	144
Authorized frequency bands	137
Authorized transmissions	129
Automatic control	119
automatic retransmission	130
Automatically controlled digital station	137
Auxiliary station	119, 132
AVVid	161
Azimuthal Map	211
B&H Sales, Inc.	160
Balun Designs	161
bandwidth	147
Bandwidth	119
Bandwidth from WPM or Baud rate	113
Barker & Williamson Antennas	162
basis and purpose of Amateur Service	119
batteries	165
Batteries America	162
Baudot	148
Beacon	119
Beacon station	132
Begali Keys	162
Bencher Keys	162
bhi Ltd.	162
Bilal Co.	162
Black Cat Systems	159
Bozak Antennas	162
BridgeCom Systems	162
broadcasting	130
Broadcasting	119
Buddipole Antennas	162
Butane powered soldering irons	164
Butternut Antennas	162
B_z	116
C.A.T.S.	162
C3F	95
Cable X-Perts, Inc.	162
cables & connectors	164
CAL-AV	162
Calculate Critical Frequency	114
Calculate Radius of Fresnel Zones	114

Call Sign Formats	41
Call Sign Prefixes (allocation -> prefix	60
Call Sign Servers	160
Call sign system	119
CallSign Software	159
Cancellation on account of the licensee's death	127
Capacitive Reactance	112
Capital Engraving	166
Carl's Electronics	162
CEPT radio amateur license	120
Certification of external RF power amplifiers	149
Chameleon Antennas	162
Champion Radio Products	162
Charlie's Electronics	162
Cheap QSL's	166
Cheapham.com	160
CHIRP	159
Circuit Specialists	162
CircuitsArchive - archive of schematics	166
Citizens Band	35
citizenship	122
Classified Ads	164
Clear Signal Products, Inc.	162
CLOVER	148
club station	122
Club Station Call Sign Administrator	125
Coaxial Cable Specifications - Attenuation	100
Coaxial Cable Specifications – Power Capacity	99
Code Quick	161
Comet Antennas/NCG meters, etc.	162
CommCat	159
Common Ham Radio Emission Types	95
Communication Concepts, Inc.	162
Communications Service Company - antennas & towers	162
Compudigital Industries	162
Computer 2100, LLC	162
ComTek Systems	162
connectors	165
Contestia	15
Contesting Software	160
Control operator	120
Control operator duties	128
Control operator required	123
Control point	120
Convert Frequency to Wavelength	112
convert prefixed values to standard values	108
convert standard values to prefixed values	108
Convert Wavelength to Frequency	112
Coordinating examination sessions	153
CQ Website News page	158
Critical Towers	162
CSCE	120
custom transformers & chokes	164
Cutting Edge Enterprises	162
CW	121, 144
Daily DX	166
Daiwa	162
Data	122
Davis Instruments	162
Davis RF	162
dB to Power Ratio	112
Dealers	160
Decibels from Power Change	112
definitions, part 97	119
Deutscher Amateur-Radio-Club	158
deviation ratio	113
Diamond Antennas	162
DigiPan	159
Dishtronix, Inc.	163
DominoEX	15
DX Atlas	159
DX cluster software for iPhone	159
DX Engineering	160
DX Maps	166
DX Store	160
DX Toolbox	159
DXLab Suite	159
DZ Company	163
E layer	116
Earth station	120, 135
Echolink	159
Education	161
e-Ham reviews of logging software	159
e-Ham.net	161
EHF	121
EIRP	12, 121
ELAD SDR	167
Elecraft	167
Electrical Values	107
Electrical Wire Run Lengths vs. Voltage Drop	98
Electromagnetic spectrum chart	165
Electron Flux	115
Electronic Formulas	112
Electronic Products Design, Inc.	163
Electronics USA	163
Element credit	152
Element standards	152
Elk Antennas	163
emergency communications	119, 127, 129, 130, 150
emergency preparedness or disaster readiness test or drill	130
Emission standards	146
emission types	93
English language	131
eQSLCard	166
Equipment Reviews	161
Equipment, parts & supplies	161
ERP	121
examination compromised	153
Examinee conduct	153
Expert Linears	163
expired license	125
External RF power amplifier	120
EZ Hang, Inc.	163
EZNEC Antenna Modeling	159
F1B	95
F3E	95
F8E	147
FAA	25, 120
false or deceptive messages	130
Family Radio Service	34
Fast Track Ham License Programs	161
FCC	120
FCC Amateur Radio Information Page	165
FCC Certified VEC's	117
FCC General Radiotelephone License Training	161
FCC modification of station license grant	127
FCC Rules & Regulations	119
FCC Universal Licensing System	165
FCC Wireless Telecommunications Bureau	165
FISTS Int'l Morse Preservation Society	158
fitted carrying cases	165
fldigi	159
FlexRadio Systems	167
footnote US270	155
Fractal Antenna Systems	163
fraudulent means	125, 153
Frequency Allocations & Band Plans	12
Frequency coordinator	120

Frequency sharing requirements	141
Front Panel Express, LLC	163
Gain of an Op-amp	113
GAP Antenna Products	163
General Mobile Radio Service	34
General station operation standards	127
Geochron, Inc.	163
Geostationary Operating Environment Satellites (GOES)	116
GHz range antennas & accessories	164
Gifts4Hams	163
Gigaparts	160
good engineering and good amateur practice	127
Green Heron Engineering LLC	163
G-TOR	148
Hagerty Radio Company	163
Half power bandwidth	113
Ham Call	160
Ham Gear	166
Ham Nation	158
Ham Radio Deluxe	159
Ham Radio Manuals	166
Ham Radio Outlet	160
Ham Services	163
HamPROs!	161
hamqsl.com	166
HamSphere	159
Harmful interference	120
Heil Sound	163
Herrman, Paul, N0BNH	116
HF	121
HF digital mode software	160
hfradio.com	166
Hip-Ham T-Shirts	166
Hy Power Antenna Company	163
Hy-Gain Antennas	163
IARP	120
IARU	158
iCluster	159
ICOM America	167
ICS Controllers	163
Image	122
immediate safety of human life	130
Impedance (Parallel)	112
Impedance (Series)	112
Impedance Matching Transformer	114
Indicator	120
Inductive Reactance	112
Industry Canada Amateur Radio page	165
Information bulletin	120
information bulletins	130
In-law	120
InnoVAntennas	163
INRAD	163
Institute for Telecommunications Services	166
Intermodulation frequency	112
International Alphabet No. 5	148
International Amateur Radio Union	159
International communications	131
International Morse code	120
International Morse Code	40
International Radio Products	163
International Telegraph Alphabet No. 2	148
interplanetary magnetic field	116
Irish Radio Transmitters Society	159
Isotron Antennas	162
ITU	29, 93, 120, 166
ITU Emission Types	93
J2A	121
J2B	95, 121, 122
J2D	93, 122
J3E	95, 150
Jetstream	163
JK Antennas	163
K4AVU Amateur Radio Products	163
KB6NU's Ham Radio Blog	158
KB9VBR Antennas	163
Kenwood	167
Kenwood renovation	162
K-index	115
KT	132
Larry's Antennas	163
LDG Electronics	163
LF	121
License term	127
Lightning Prevention Systems, Inc	163
Line A	23, 120
Local control	120
Log periodic antennas	163, 165
Log4OM	159
Logbook of The World	160
Logger 32	160
Lowest Usable Frequency	115
M2 Antenna Systems, Inc.	163
MacLoggerDX	160
Mailing address	126
manned spacecraft	130
ManualMan	166
Marine VHF Frequencies	37
Mastrant	164
material compensation	130
Maximum Usable Frequency	115
MCW	122
Message forwarding system	120, 136
Metric Electronic Values	108
MF	121
MFJ Enterprises	161
MFSK16	15
Mica Capacitor Color Code	110
military recreation station	122
minimum transmitter power necessary	148
Mitchell Electronics	164
MMSSTV	160
MMTTY	160
modulation index	147
Modulation Index	113
monetary or other consideration	153
Monitoring and analysis tools	162
Morse Code	40
Morse code decoding software	160
Morse Express	164
MT63	15
MTC	161
MUF	115
MultiPSK	160
MURS (Multi-Use Radio Service)	35
music	130
N1MMlogger	160
N3JFP Contesting Software	160
N3ZN Keys, LLC	164
NASA	166
National Aeronautics and Space Administration	130
National Hurricane Center	166
National Institute of Standards & Technology	166
National Oceanic & Atmospheric Administration	166
National Radio Astronomy Observatory	132
National Radio Quiet Zone	120, 132
National RF, Inc.	164
National Telecommunications and Information Administration	166

National Weather Service	166
NATO Phonetic Alphabet	40
NCDXF Beacon Stations	39
Nemal Electronics	164
New Zealand Association of Radio Transmitters	159
news gathering	130
NOAA	115
NOAA Space Environment Center	167
NOAA Storm Prediction Center	166
NOAA Weather Radio Frequencies	35
Norms Fabrication	164
number of examinees	152
NYQSL	166
obscene or indecent words or language	130
obscuring meaning (of message)	130
OET Bulletin Number 65	124
Official Time	166
Ohm's Law	111
Old Heathkit Parts	164
Olivia	15
OLYMPIX	164
one-way communications	130
Online Cables	164
Open Repeater Project	160
Operation during a disaster	150
Operator license grant	123
operator/primary station license grant	122
Optimum "Random" Wire Antenna Lengths	104
PACTOR	15, 148
Palomar Engineers	164
Palstar	164
Parallel Resistance, Parallel Inductance, and Series Capacitance	112
Peak Voltage from Peak-to-Peak	113
Peak Voltage from RMS Voltage	113
pecuniary interest	119, 130
PEP	121
Percent Change of Power Increase from Decibels	112
Percent Power Remaining After dB Change	112
Personalized Map Company	166
Peter W. Dahl	164
Phase Angle	113
Phone	122
phonetic alphabet	40, 131
Photo QSL's	166
Physician	120
Pixel Satellite Radio	164
Places Where the Amateur Service is Regulated by the FCC	154
POLAR Electric	160
Polyphaser Corporation	164
Power Dissipation of a Series Connected Linear Voltage Regulator	113
Power Factor	113
Power Loss from SWR	105
Power Ratio to dB	112
Powerpoles®	164
PowerWerx	164
preciseRF	164
Precision Resistors	164
Preparing an examination	152
printable county maps	166
program production	130
Prohibited transmissions	130
Prolog	160
propagation information	130
Propagation Information	166
Proplab-Pro	160
ProPrints	166
Prosigns	88
Prosigns For CW Net Use	91
protection of property	130
Proton Flux	116
Pulse	122
Q of a Parallel Resonant Circuit	112
Q of a Series Resonant Circuit	112
Q Signals	86
QN Signals For CW Net Use	90
QRO Technologies, Inc.	164
QRZ.com	158, 160
QTH.com	164
Qualifying Examination Systems	151
Qualifying for an amateur operator license	151
Quarter Century Wireless Assn.	159
Question pools	154
R & L Electronics, Inc	164
R3E	95, 150
RACES	121, 127, 129, 150, 156
Radio amateur civil emergency service	150
Radio Amateur Satellite Corp.	159
Radio Amateurs of Canada	159
Radio Manufacturers	167
Radio Society of Great Britain	159
Radio-Electronics.com	161
Reciprocal operating authority	129
reciprocal operation station identification	132
Reimbursement for expenses	154
Remote control	121
Renew Your License	156
Repeater	121
Repeater Listings	167
Repeater station	133
Repeaterbook	167
Replacement license grant document	127
Rescue Tape	164
Resistor Color Codes	109
Resonant Frequency of a Circuit	112
Restricted operation	132
Restrictions on station location	123
retransmission (of programs or signals)	130
RF	121
RF Connection	164
RF Exposure Limits	106
RF Gear 2 Go	164
RF Parts Company	164
RF Spectrum Allocation Chart PDF	166
RF Toolbox	160
RFI Filters	164
RigExpert	164
RMS Voltage from Peak Voltage	113
Rotating Tower Systems	164
Rotor parts	162
RT Systems	160
RTTY	122
RTTY and data emission codes	148
Safety of life and protection of property	150
sale or trade of apparatus normally used in an amateur station	130
Sample Station Log Page	210
Schematic Symbols	101
sequential call sign system	125
Sequential call sign system	119
Series Resistance, Series Inductance, and Parallel Capacitance	112
SFI	115
SHF	121
Smith Chart	118
solar flare	116
Solar Flux Index	115
solar wind	115
Solar Wind	116

Solder-it	164
Space station	121, 133
Space telecommand station	136
Space telemetry	121
space weather	115, 116
SpaceWeather	167
Special event call sign system	119
Spread spectrum	122
Spurious emission	121
spurious emissions	146
SS	122
SS emission types	148
Standards for certification of external RF power amplifiers	149
station aboard an aircraft	123
Station antenna structures	124
Station control	129
Station identification	131
Station in distress	150
station license	122
Station licensee responsibilities	127
Stations aboard ships or aircraft	123
SteppIR Antennas Inc.	164
Sun's magnetic field	116
sunspot number	115
sunspots	115
Super Antenna	164
Susceptance	113
TAPR	159
Tarheel Antennas, Inc.	164
teacher	130
Telecommand	121
Telecommand of an amateur station	136
Telecommand of model craft	136
Telecommand station	121
Telemetry	121, 136
Tennadyne, L.L.C.	165
Ten-Tec, Inc.	165
Ten-X International Net	159
Test	122
Third party communications	121, 130
Third Party Operating Agreements	73
Time Constant of a Capacitor	113
Time Constant of an Inductor	113
Timewave Technology	165
Tower	164, 165
Tower building book	165
Tower parts	164
Towers	162
Transformers	164
Transmitter power standards	148
TrueTTY	160
T-Shirts	166
Tucson Amateur Packet Radio	159
Turns on a Coil for a Desired Inductance	113
UHF	121
ULS	121, 156
Uniden America Corporation	165
Univ. of Arkansas / Little Rock	160
Universal Radio, Inc.	161
unspecified digital code	148
US Call Sign Regions	41
US Tower Corporation	165
Vanity Call Sign	157
Vanity call sign info	160
Vanity call sign system	119
VE	121, 125, 152, 154
VE session manager requirements	153
VEC	117, 121, 152, 153, 154, 156
VEC Regions	155
Ventenna	165
VHF	121
Vibroplex	165
Vintage Manuals, Inc.	166
Voltage at the Secondary Winding of a Transformer Based on Turns (N)	113
Volunteer Examiner (license requirement)	10, 11
W&W Manufacturing Co	165
W2IHY Technologies	165
W5YI	161
Waterproof log books	161
WaveNode	165
WD7S Productions	165
Weather Defender	165
weather forecast	130
Weather tracking software	165
Web Links	158
West Mountain Radio	165
Western Case	165
Wire Gauge	96
Wire Size	96
Wire Types	97
Wired Communications	165
Wireman	165
WK5R Circuit Boards	165
Wolf number	115
World Voltages	106
WSJT Suite	160
WSPRNET	167
WWV	166
Xmatch Antenna Tuner	165
X-ray Flux	116
Yaesu USA	167
YL Radio League	159
YLRL	159
Your Frequency Privileges	10

Sample Station Log Page

Call Sign _____ **Station Log** **Page** _____

Date	Time (UTC) On	Time (UTC) Off	Freq	Mode	Power	Call Sign	QTH	R/S/T Sent	R/S/T RCVD	Comments

Azimuthal Map

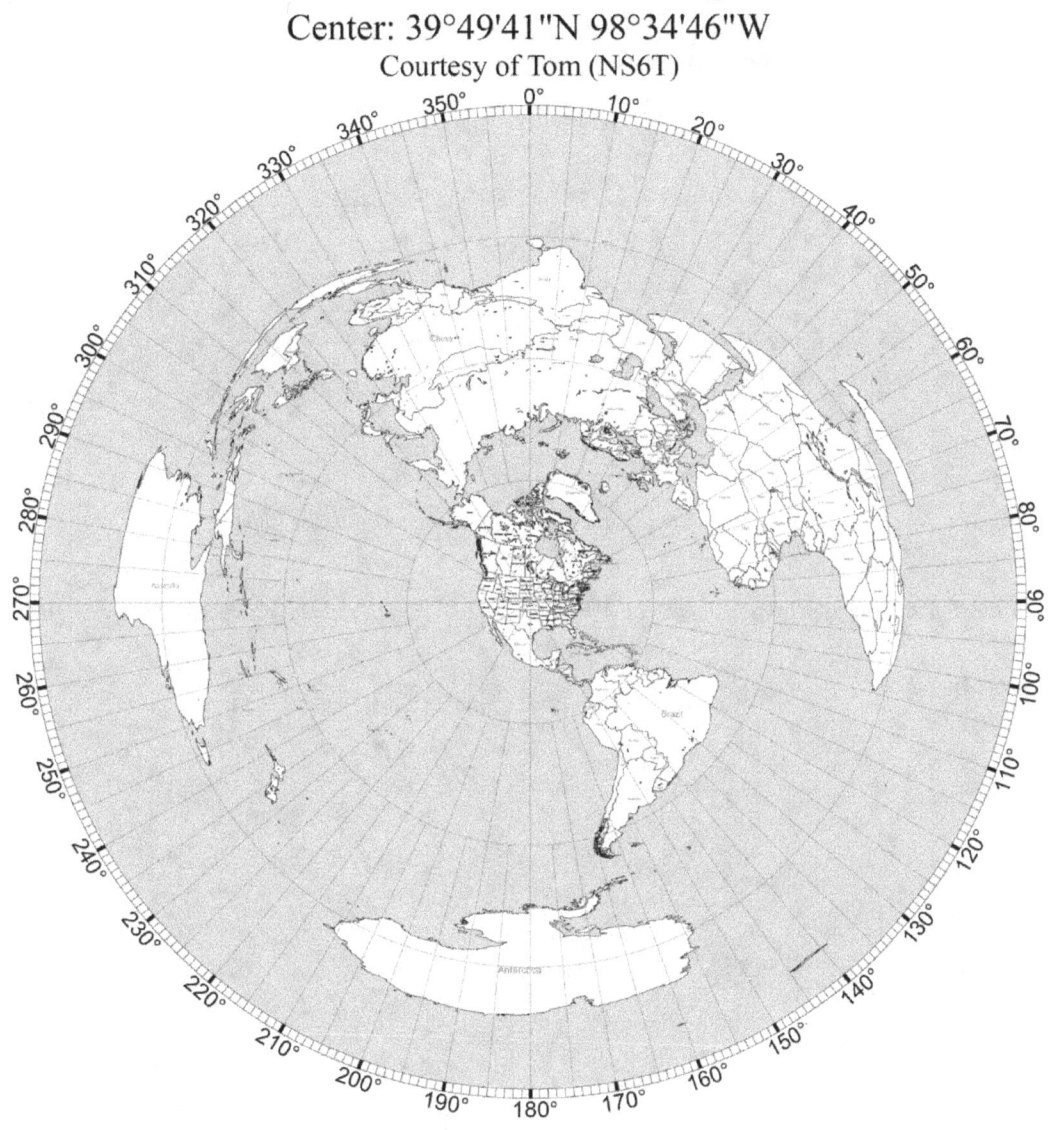

By the Same Author

The Fast Track to Getting Started in Ham Radio

The Fast Track to Your Technician Class Ham Radio License
The Fast Track to Mastering Technician Class Ham Radio Math

The Fast Track to Your General Class Ham Radio License
The Fast Track to Mastering General Class Ham Radio Math

The Fast Track to Your Extra Class Ham Radio License
The Fast Track to Mastering Extra Class Ham Radio Math

The Fast Track to Understanding Ham Radio Propagation
The Fast Track to (Finally!) Getting on the Air with Ham Radio

The Independent Author's Guide to Audiobook Production: A Professional Narrator's Secrets for Success on ACX

Come visit our booth at major hamfests!

www.ingramcontent.com/pod-product-compliance
Lightning Source LLC
Chambersburg PA
CBHW080454220526
45465CB00006B/2268